Happy Christmas &
New Year 2016!

-Rihtik

THE COMPLETE GUIDE TO HOW THINGS WORK

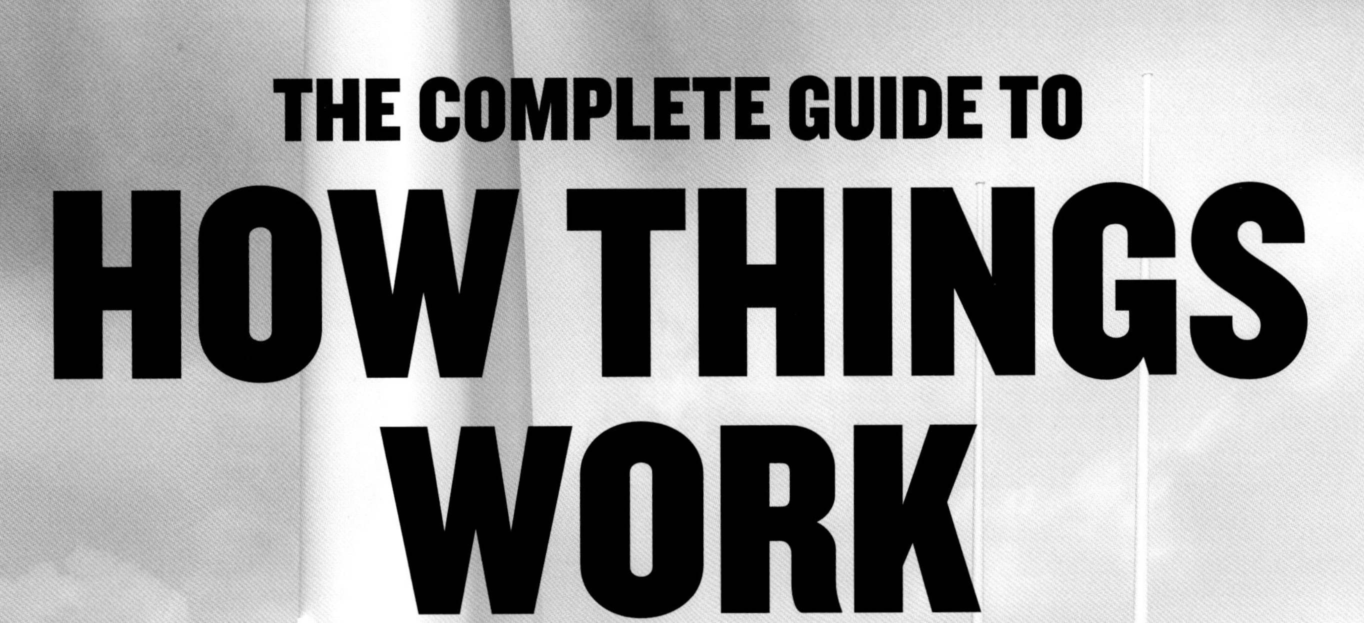

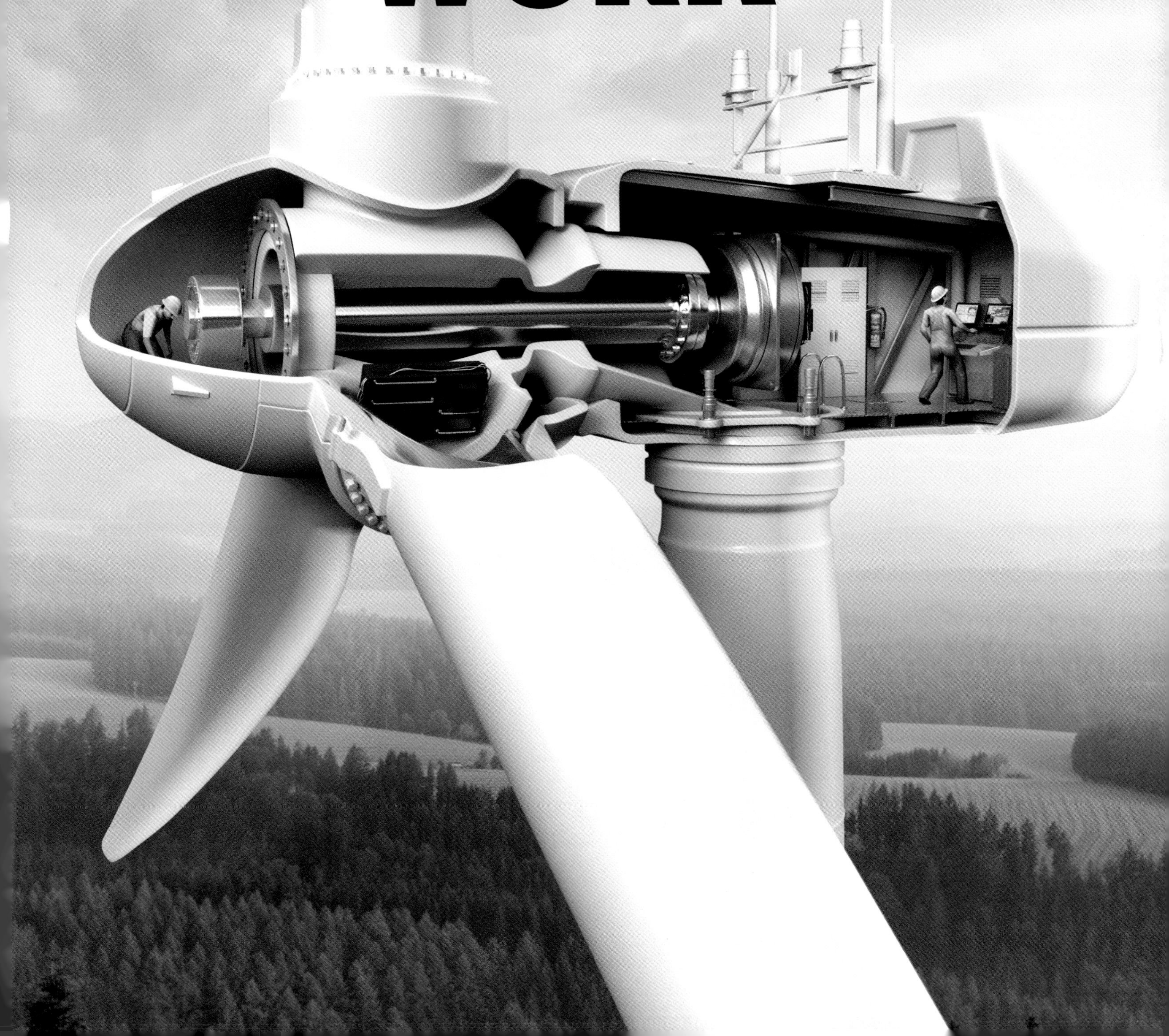

An Imprint of Sterling Publishing
1166 Avenue of The Americas
New York, NY, 10036

This 2015 edition published by Sandy Creek.

ISBN 978-1-4351-6163-4

Manufactured in China

Lot #:

2 4 6 8 10 9 7 5 3 1

08/15

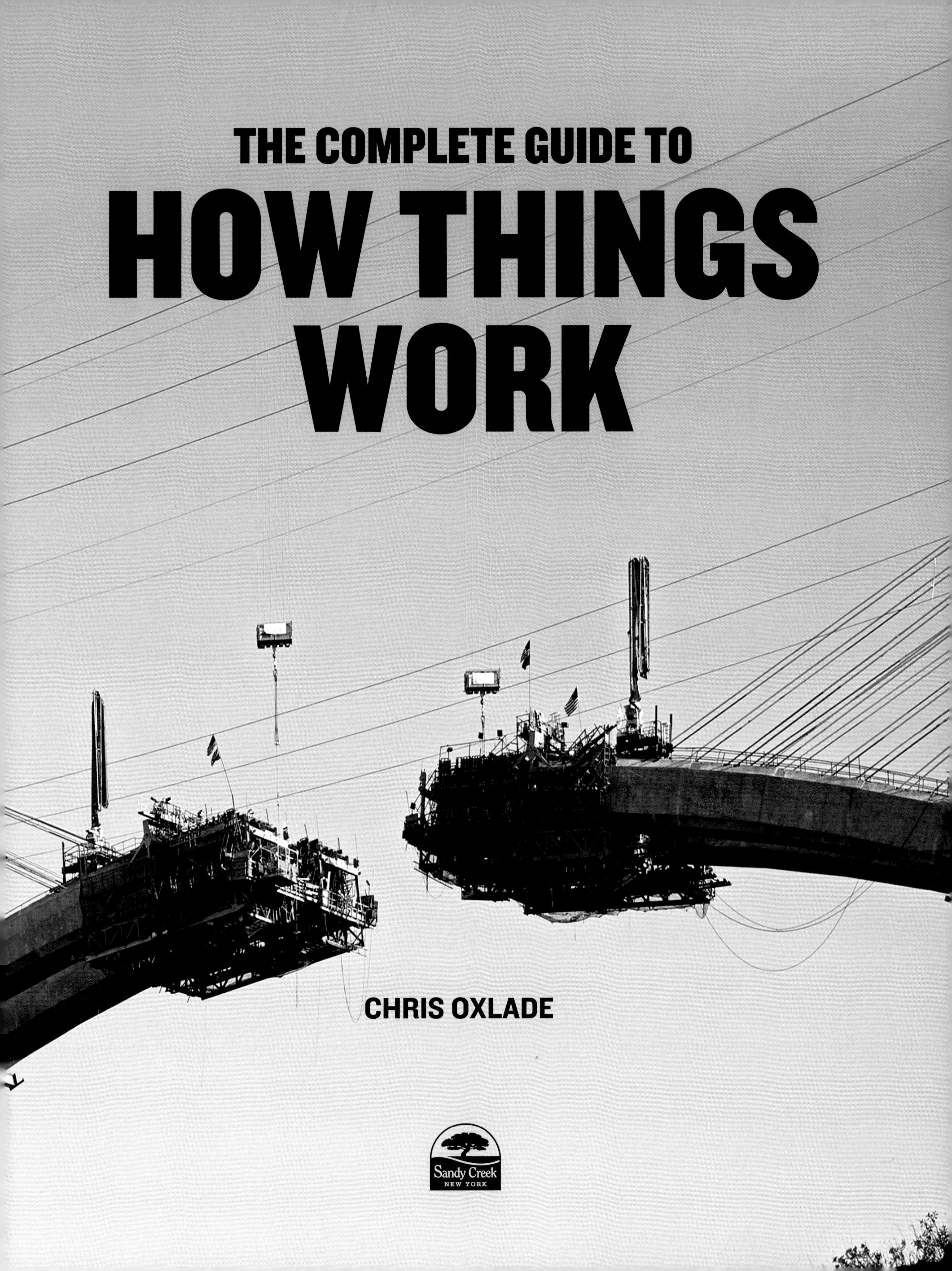

THE COMPLETE GUIDE TO
HOW THINGS WORK

CHRIS OXLADE

Sandy Creek
NEW YORK

CONTENTS

DID YOU KNOW?

Words in **bold** are explained in the Glossary on page 140.

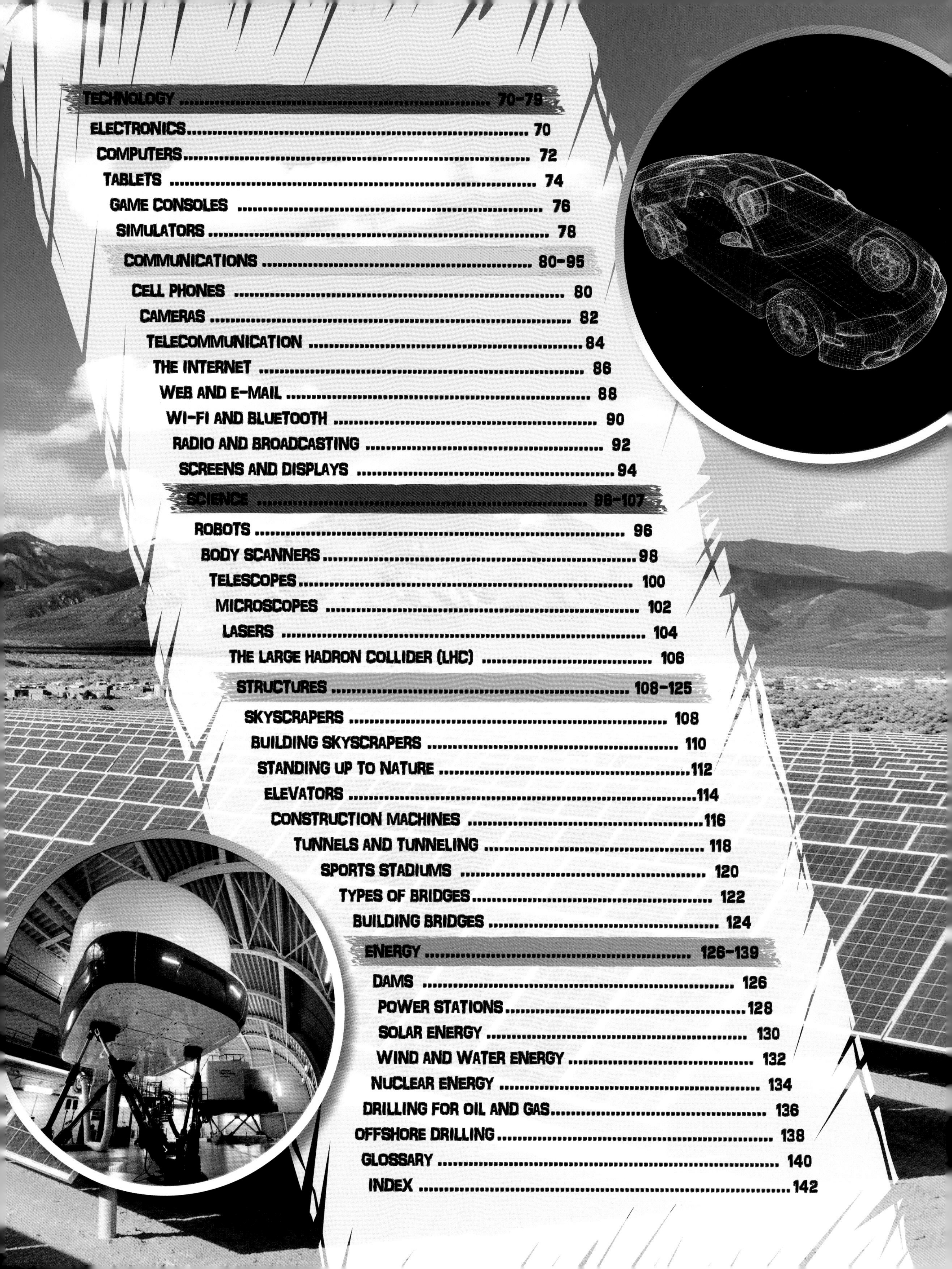

ENGINES

All sorts of machines, from cars to lawn mowers to giant cargo ships, have engines. We put **fuel,** such as gasoline or diesel, into the engine, and the engine turns the fuel into energy to make the parts of the machine move.

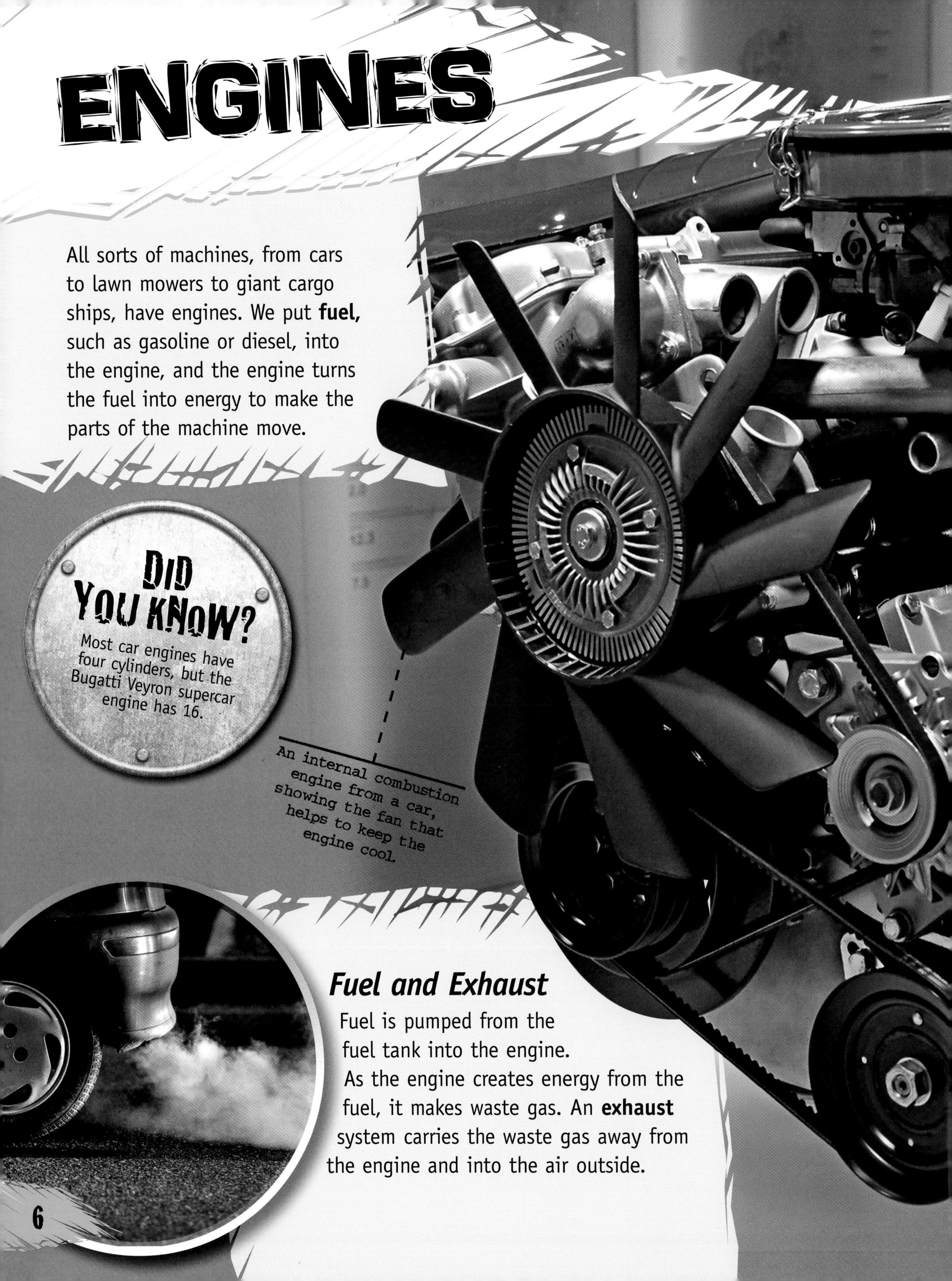

DID YOU KNOW?

Most car engines have four cylinders, but the Bugatti Veyron supercar engine has 16.

An internal combustion engine from a car, showing the fan that helps to keep the engine cool.

Fuel and Exhaust

Fuel is pumped from the fuel tank into the engine. As the engine creates energy from the fuel, it makes waste gas. An **exhaust** system carries the waste gas away from the engine and into the air outside.

Parts of an Engine

The **internal combustion engine** is the most common type of engine. Fuel burns in cylinders inside the engine. Pistons slide up and down inside the cylinders. As the pistons move, they spin the crankshaft, the part of the engine that powers machinery.

Moving the Pistons

Each piston in an internal combustion engine moves up and down in its cylinder. It goes through four stages, called a four-stroke cycle, and repeats this cycle again and again. These diagrams show the cycle in an internal combustion engine powered by gasoline.

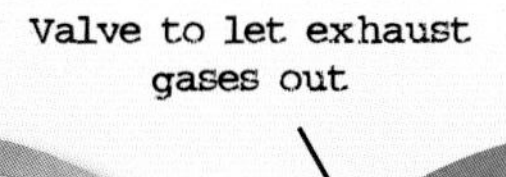

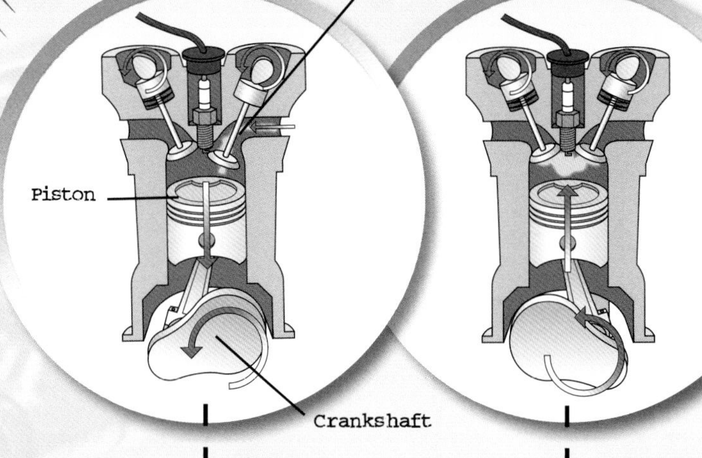

1. INTAKE STROKE: A valve opens to let the fuel and air mixture into the cylinder. The piston moves down, sucking the fuel and air mixture into the cylinder.

2. COMPRESSION STROKE: The valves close. The piston moves up, squeezing the fuel and air mixture into the top of the cylinder.

3. POWER STROKE: A spark ignites the fuel and the explosion pushes the piston down. The piston pushes the crankshaft around.

4. EXHAUST STROKE: The exhaust valve opens to let waste gas out. The piston moves up, pushing waste gases out of the cylinder.

CARS

All cars, from small city cars to limousines, have similar parts. The main part of any car is its body, which is made of metal panels **welded** together to make a stiff structure called the frame. All other parts of the car are connected to the frame.

DID YOU KNOW?

The average family car is made from more than 30,000 parts.

Car Parts

Most cars are powered by an internal combustion engine (see pages 6-7). The engine turns the front wheels, the rear wheels, or in a four-wheel-drive car, all the wheels. **Suspension** lets the wheels move up and down as the car goes over bumps, and tires grip the road. A steering system swivels the front wheels to make the car turn left or right.

Driver Controls

A driver can make a car speed up or slow down using foot pedals. **Gears** allow the car to start and travel at different speeds. In an automatic car, the car chooses the gears. In a manual car, the driver selects the gears. The steering wheel makes the car turn left or right. Most cars have power steering that helps the driver to swivel the wheels. There are lots of switches for controlling lights, windshield wipers, and other devices in the car.

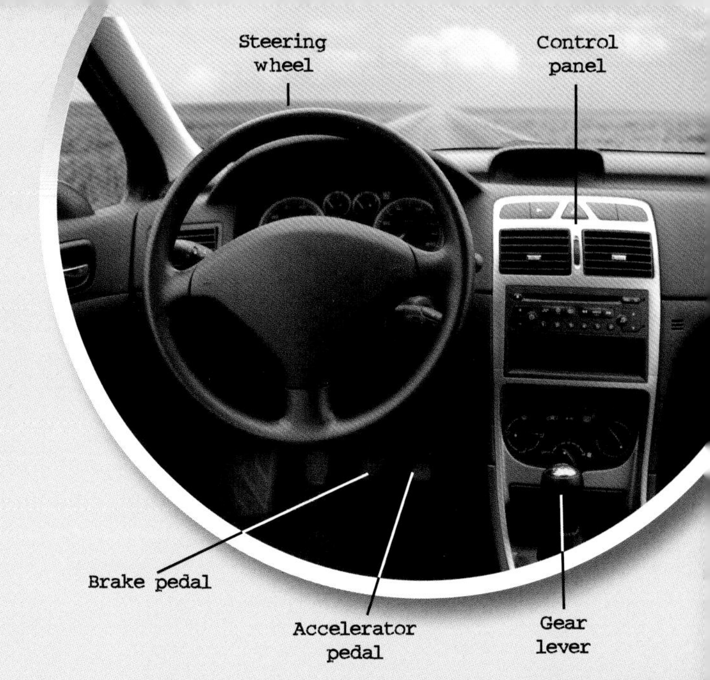

This cutaway of a sports car shows how all the parts of a car fit together inside.

Driverless Cars

Many cars help drivers automatically by controlling the car's speed, or by braking if they get too close to other cars. Engineers are experimenting with cars that drive completely automatically, without any help from a driver.

An experimental driverless car made by Google.

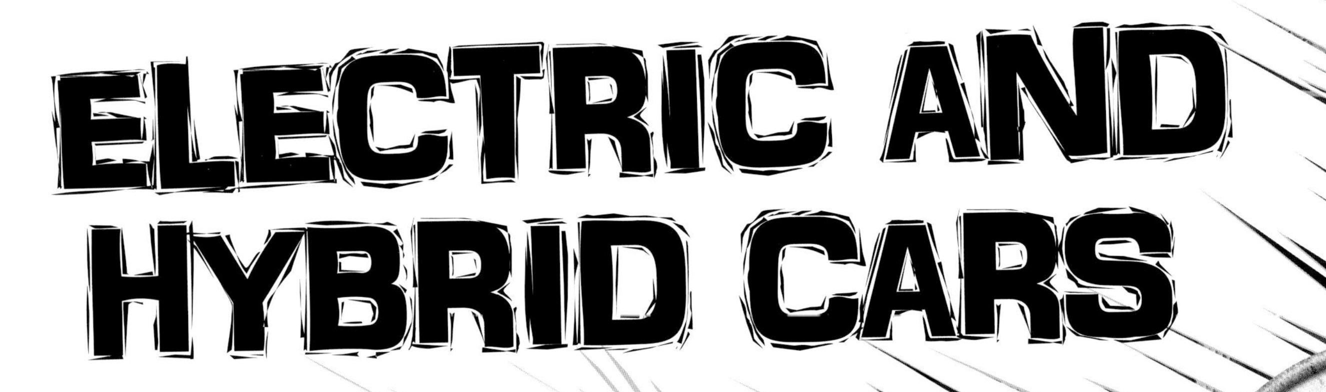

ELECTRIC AND HYBRID CARS

Electric cars are powered by electric motors. They are almost silent and don't produce the exhaust gases, such as carbon dioxide, that cars with internal combustion engines do.

DID YOU KNOW?

The Detroit Electric SP:01 is the fastest electric car, with a top speed of 155 miles per hour.

Plugging In

In an electric car, the electricity that works the motors comes from batteries. The driver controls the car's speed with a pedal that controls how much electricity flows from the batteries to the motor. The batteries must be recharged regularly to keep the car going. In an electric car, there is enough electricity in fully charged batteries for about 100 miles of driving.

This electric car is plugged in at a recharging station.

Battery Banks

The batteries are normally in the back or under the floor of an electric car. Engineers are making batteries that are lighter in weight and store more electricity, so that electric cars can travel farther and faster.

Electric cars have a bank of batteries all connected to one another.

Hybrid Cars

A **hybrid** car has both a gasoline engine and an electric motor. The car automatically uses the electric motor when the car starts to move, and at low speeds. It uses the engine at higher speeds, when the engine is also used to recharge the batteries. Hybrid cars are more efficient than normal cars, and can keep going longer than electric cars.

Recharging station

Recharging cable plugged into car.

The electrical and mechanical parts under the hood of a modern hybrid car.

FAST CARS

Fast cars include sports cars, racing cars, and record-breaking cars. They all have special features that make them faster than normal cars, such as super-powerful engines, **streamlined** bodies, and wide tires.

A streamlined shape allows this car to cut through the air easily.

Streamlining

Cars must push the air out of the way as they speed along. Fast cars are designed with a smooth, streamlined shape so that the air slips easily over and around them. The faster a car goes, the more important streamlining becomes.

DID YOU KNOW?

The current fastest car is the Thrust SSC. In 1997, it reached a record-breaking 763 miles per hour—faster than the speed of sound travels.

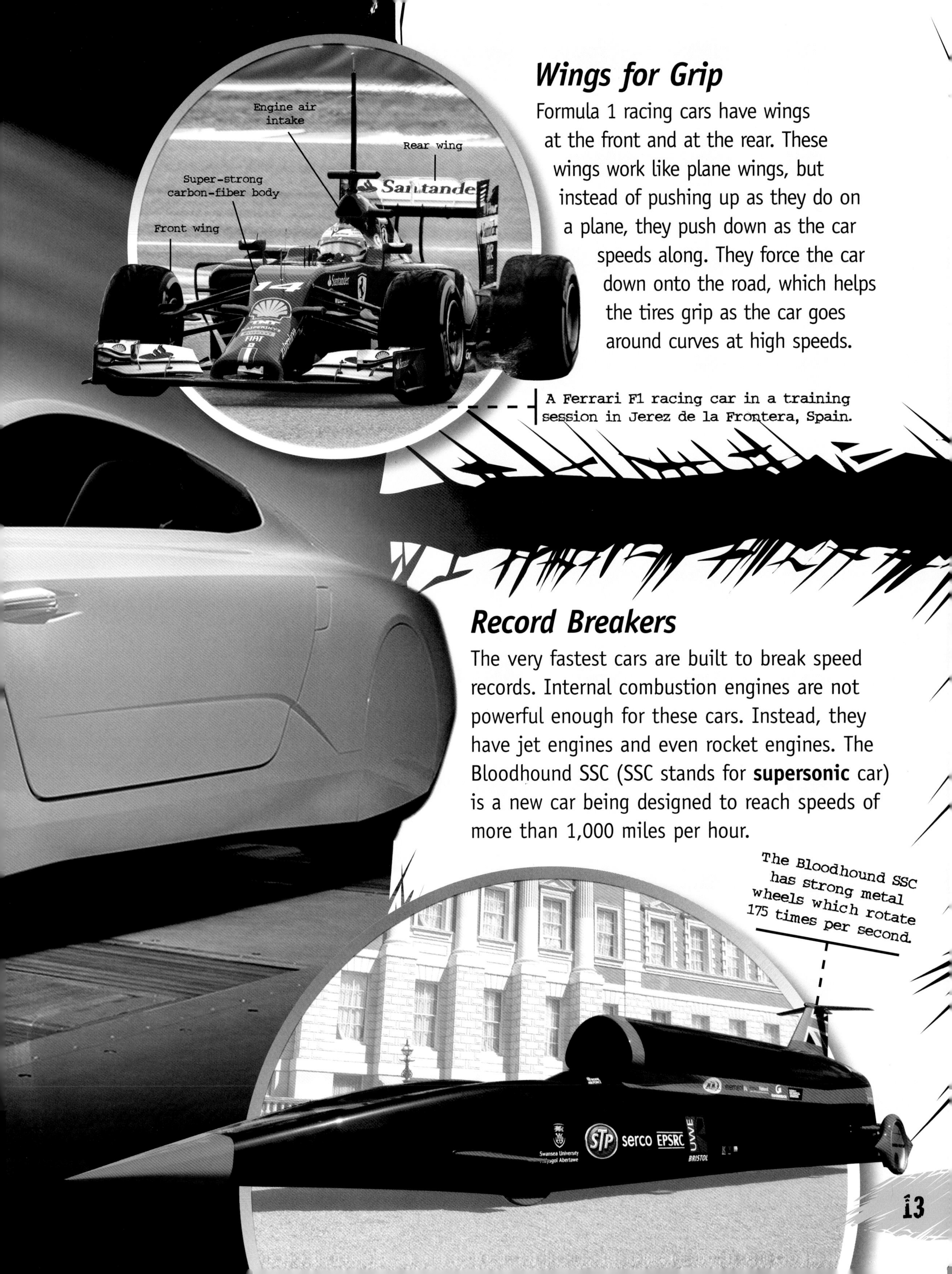

Wings for Grip

Formula 1 racing cars have wings at the front and at the rear. These wings work like plane wings, but instead of pushing up as they do on a plane, they push down as the car speeds along. They force the car down onto the road, which helps the tires grip as the car goes around curves at high speeds.

A Ferrari F1 racing car in a training session in Jerez de la Frontera, Spain.

Record Breakers

The very fastest cars are built to break speed records. Internal combustion engines are not powerful enough for these cars. Instead, they have jet engines and even rocket engines. The Bloodhound SSC (SSC stands for **supersonic** car) is a new car being designed to reach speeds of more than 1,000 miles per hour.

The Bloodhound SSC has strong metal wheels which rotate 175 times per second.

TRUCKS

Trucks are tough vehicles that carry heavy loads, often on long journeys. There are two main types of trucks: semitrailer trucks and rigid trucks. There are general-purpose trucks and specialized trucks, such as tankers, dump trucks, and refuse trucks.

DID YOU KNOW?

A large truck engine produces around 750 **horsepower**—about seven times as much power as the engine in a family car.

Semitrailer Truck

A semitrailer is made up of a tractor unit and a trailer that the tractor unit pulls along. The tractor unit has a very powerful **diesel engine**, and the driver has as many as 18 gears to choose from. At the rear of the tractor unit is a big metal plate, called the fifth wheel, which a trailer connects to. The semitrailer's wide wheels help to spread the heavy weight of the truck's load.

Streamlined cab

Tractor unit

Diesel engine under cab

Fifth wheel

Fuel tank

Large, wide wheels

semitrailer truck and driver's cab.

In this big semitrailer truck, the diesel engine is in front of the driver's cab.

Rigid Truck

Unlike a semitrailer truck, all the parts of a rigid truck are attached to one unit—a strong metal frame called a chassis. The engine is always at the front of the chassis. It turns a driveshaft that turns the rear wheels, which drive the truck along. The cab is built on top.

A Ford 150 Raptor truck chassis displayed on its side.

Going Off-road

Fire trucks, forestry trucks, construction trucks, and military trucks often have to drive across rough and muddy ground or through rivers. They have special suspension that allows their wheels to move up and down over rocks, and chunky tires and extra-low gears for going up steep hills.

A military truck with high suspension crosses a river.

BICYCLES AND MOTORCYCLES

Bicycles are machines powered by humans. Motorcycles are really just bicycles with an engine added to drive them along. Machines such as quad bikes and snowmobiles work in a similar way to motorcycles.

Kymco Venox 250 motorcycle.

Frame

Engine

Rounded tires for leaning over on corners

Motorcycle Parts

The main part of a motorcycle is a strong but lightweight metal frame. All the other parts are attached to the frame. The engine is always at the bottom of the frame. The rear wheel is turned by a chain connected to the engine, just as the rear wheel of a bicycle is turned by pedals. The rider uses pedals and levers to control the bike.

Track Cycles

Indoor track cyclists ride specialized bicycles built for speed. They have frames made of **carbon fiber**, which is a strong, lightweight material. All the parts are designed to be smooth and streamlined, including the wheels.

Handles
Wheels
Frame
A track cyclist can reach speeds of 50 miles per hour on a special track bike.
Suspension
DID YOU KNOW?
The fastest electric motorcycle is the Lightning LS-218, with a top speed of 218 miles per hour.
Snowmobiles are powered by internal combustion engines.
Track
Skids

Bikes on Snow

A snowmobile (also called a skidoo) is like a motorcycle with the front wheel replaced by skids that work like skis. The rear wheel is replaced by a moving track. The rider turns the skids from side to side to steer. The track grips the snow as it pushes the snowmobile forward.

TANKS

DID YOU KNOW?

The gun of an A1 Abrams tank can hit a target 2.5 miles away 19 times out of 20.

A tank is a heavy fighting vehicle with a powerful gun. The crew (normally made up of a commander, a gun aimer, a gun loader, and a driver) travel inside the tank, protected by the tank's thick armor.

This is an M1 Abrams battle tank, one of the world's largest tanks.

The turret revolves to aim the gun at targets.
Tracks
Armor

Battle Tanks

Battle tanks are the heaviest tanks with the biggest guns. A battle tank has armor called **composite** armor to protect it against exploding anti-tank shells. The armor is made from layers of metal, plastics, and a super strong material called **Kevlar**, which is the same strength as a sheet of steel that is five feet thick.

Tracks for Grip

Tanks have two metal tracks called caterpillar tracks. On rough, uneven ground, tracks give much better grip than wheels, and they help to stop the heavy tanks from sinking into mud. The driver can control each track separately and can make the tank spin around quickly by applying the brakes on one of the tracks.

A tank climbing up an obstacle using its tracks.

The hottest parts of another tank look brightest through a tank's thermal-imaging sight.

Taking Aim

The tank gun has a laser **range** finder to calculate the distance it has to fire to reach a target, and a thermal-imaging sight to see targets at night. Once the gunner has selected a target, a computer swivels the turret and adjusts the angle of the gun automatically before firing.

TRAINS

Power cable
Pantograph
The driver of an electric train controls how much electricity is fed to the motors to manage the train's speed.

Trains are vehicles with metal wheels that run on railway tracks. There are several different types of power used for modern trains. The main types are electric, diesel, and diesel-electric.

Electric Trains

Electric trains are driven by electric motors that turn the wheels. An electric **locomotive** collects electricity from an overhead cable through a pantograph (electric current collector) attached to the top of the train. High-speed electric trains have power carriages instead of a separate locomotive.

DID YOU KNOW?

In 2015, Japan's maglev train reached a record speed of 375 miles per hour.

Diesel Trains

Diesel locomotives are powered by enormous diesel engines, which drive the wheels of the train. Many trains are also pulled by diesel-electric locomotives. The diesel engine turns an electricity **generator** to make electricity, and the electricity drives electric motors that turn the wheels.

A diesel-electric locomotive pulls long freight trains in Alaska.

Frame with wheels and motors called a bogie

Maglev Trains

Maglev is short for magnetic levitation. Powerful magnets in the track and in the train itself make a maglev train hover just above the track. More magnets make the train move forward. Because maglev trains don't touch the track, they run very smoothly and at very high speeds.

This Japanese high-speed maglev is still at the experimental stage and doesn't carry passengers yet.

TRAFFIC AND TRAIN CONTROL

Traffic on our roads needs to be controlled to keep us safe and to keep vehicles flowing freely. Cars and other road vehicles are controlled by road traffic signals. Trains are also controlled to keep them at safe distances from one another. This is done at a railway control center.

Railway Control Center

A railway network has many lines that cross and join one another at junctions. Controllers at a railway control center watch trains on electronic maps of the network. Sensors on the tracks show them where the trains are and they use this **data** to control the signals.

Track Signals

Train drivers watch signals alongside or above the tracks. Different colors and patterns of lights on the signals tell the drivers what speed they are allowed to travel, or whether the track ahead is clear or if it is blocked.

A railway control center uses data from sensors to monitor the traffic flow.

Railway Crossing

At a railway crossing, sensors on the railway track detect trains approaching the crossing. An automatic system switches on flashing lights and closes gates to stop traffic well before the train hurtles by.

Did You Know?

The USA's first electric traffic lights were installed at a road junction in Cleveland, Ohio, in 1914.

A train passing under a signal gantry. Red lights stop trains from going in the opposite direction.

Red lights warn when a train is approaching.

Road Traffic Signals

At a road junction the sequence of lights (usually red, yellow, and green) is controlled by electronics in a box at the junction. At some junctions there are sensors above the street or under the road that can tell when cars are waiting to go, and the sensors adjust the lights to keep the traffic flowing freely. Human controllers at a traffic management center can override the signals to clear congestion.

At a busy junction, traffic lights and traffic police help guide the cars.

SUBWAYS

DID YOU KNOW?
The USA's first subway opened in Boston in 1897. More than 100,000 people rode the new subway trains on the opening day.

A subway is a railway system underneath the city streets. Subway trains run through tunnels between underground stations. Subways move passengers around a busy city quickly because the trains avoid heavy traffic.

New York City Subway's A train travels more than 31 miles from start to finish.

Subway Trains

All subway trains, such as this one that runs along Eighth Avenue in New York City, are powered by electric motors, as dangerous fumes from engines would pollute the air in the tunnels. A computerized train control system keeps trains on the same line at a safe distance from one another. Some systems, such as the Copenhagen Metro in Denmark, are completely automated, with driverless trains.

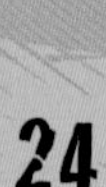

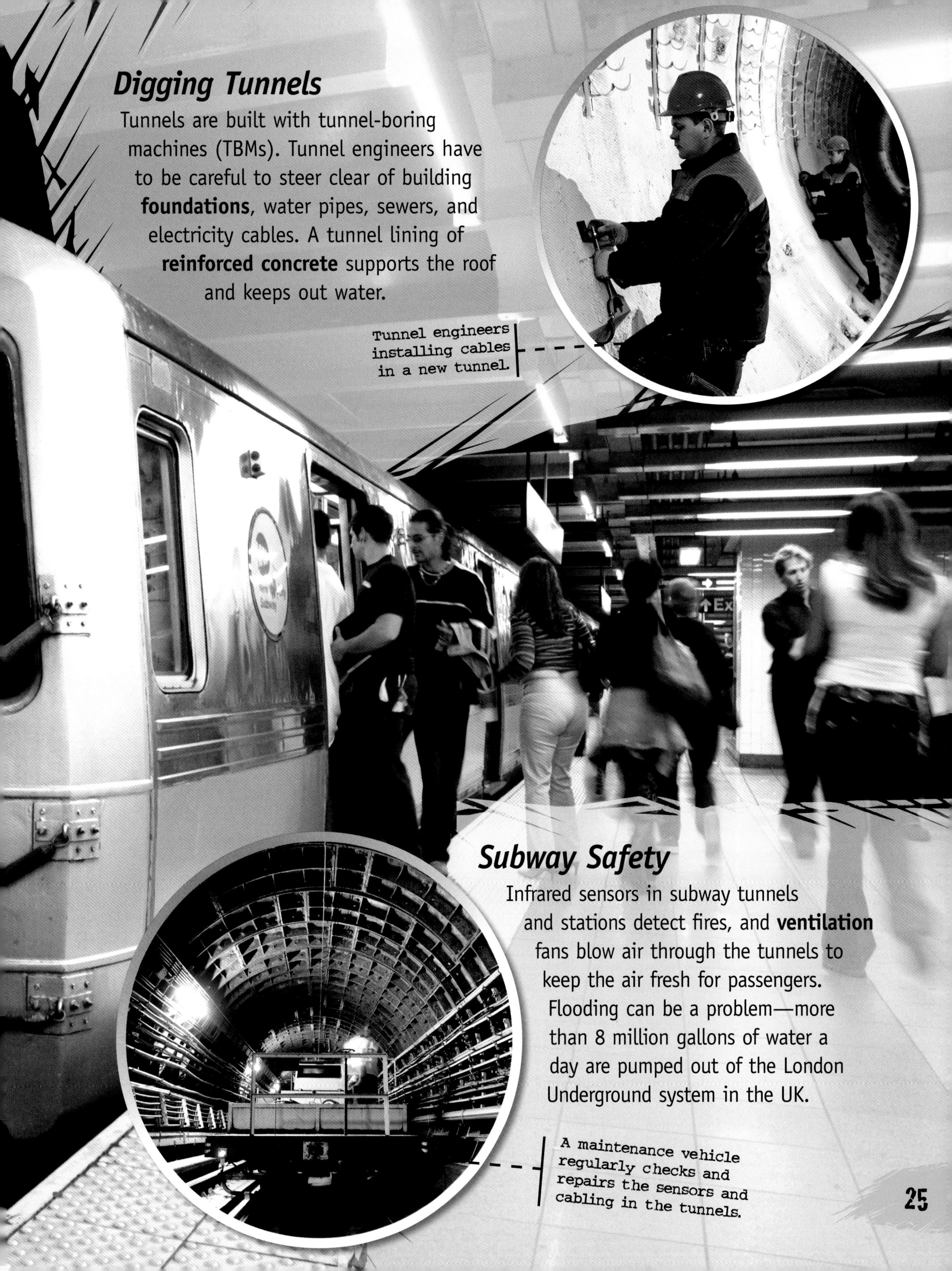

Digging Tunnels

Tunnels are built with tunnel-boring machines (TBMs). Tunnel engineers have to be careful to steer clear of building **foundations**, water pipes, sewers, and electricity cables. A tunnel lining of **reinforced concrete** supports the roof and keeps out water.

Tunnel engineers installing cables in a new tunnel.

Subway Safety

Infrared sensors in subway tunnels and stations detect fires, and **ventilation** fans blow air through the tunnels to keep the air fresh for passengers. Flooding can be a problem—more than 8 million gallons of water a day are pumped out of the London Underground system in the UK.

A maintenance vehicle regularly checks and repairs the sensors and cabling in the tunnels.

CHANNEL TUNNEL

The Channel Tunnel runs between Britain and France. For most of its length, it runs under the water of the English Channel. Digging began in 1987 and the first trains ran through it in 1994. High-speed passenger trains and vehicle shuttle trains travel through the Channel Tunnel.

DID YOU KNOW?

The Channel Tunnel is 31 miles long. 24 miles of it is under the sea, making it the longest undersea tunnel in the world.

Three Tunnels

The Channel Tunnel is made up of three tunnels: two railway tunnels that are 25 feet wide and a 52 foot long service tunnel that runs between them. Workers use the service tunnel to make repairs. It can also be used as an escape route in an emergency, such as a fire in the railway tunnel.

A Eurostar high speed train exits the French portal at Calais following the 35 minute journey through the Channel Tunnel.

Giant Caverns

In the Channel Tunnel, there are two giant underground caverns called crossovers. This is where the two main tunnels meet to let trains cross between the tracks. This happens if one part of the tunnel has to be closed for repairs.

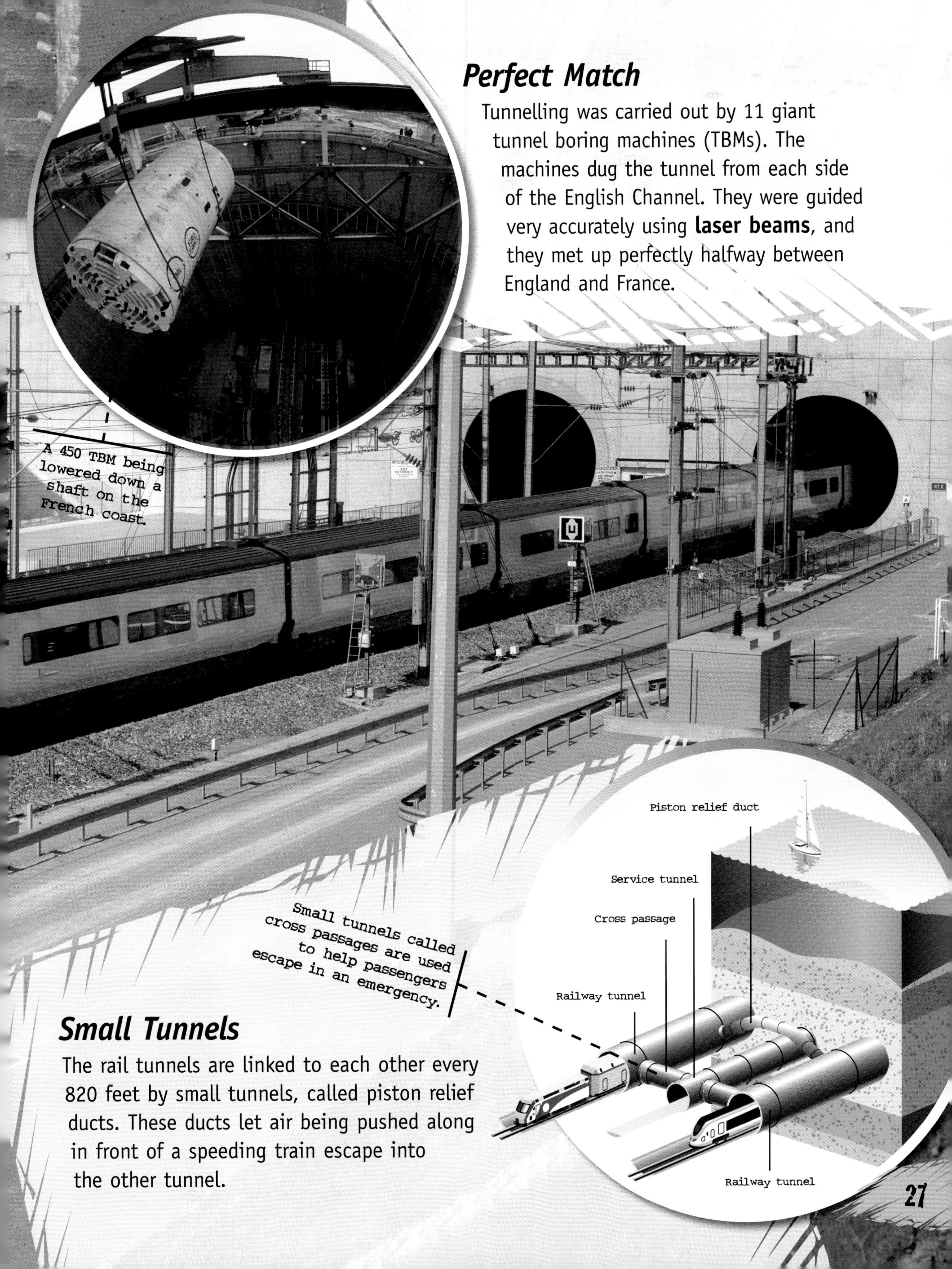

A 450 TBM being lowered down a shaft on the French coast.
Small tunnels called cross passages are used to help passengers escape in an emergency.
Piston relief duct
Service tunnel
Cross passage
Railway tunnel
Railway tunnel

Perfect Match

Tunnelling was carried out by 11 giant tunnel boring machines (TBMs). The machines dug the tunnel from each side of the English Channel. They were guided very accurately using **laser beams**, and they met up perfectly halfway between England and France.

Small Tunnels

The rail tunnels are linked to each other every 820 feet by small tunnels, called piston relief ducts. These ducts let air being pushed along in front of a speeding train escape into the other tunnel.

SHIPS

The V-shaped bow of a cruise ship's hull pushes the water aside as the ship moves along.

A ship is a large vessel designed to travel across seas, oceans, rivers, and lakes. Ships are used for leisure and holidays, and to transport goods and people. Smaller vessels are called boats rather than ships. All ships and boats float because their hulls (the frame of the ship) push water aside as the vessels sit in the water. The water pushes back with a force called upthrust.

Parts of a Ship

Most modern ships are made of steel. The hull is a steel structure that is hollow inside and covered with steel plates. The hull is shaped to make the ship stable and allow it to slip easily through the water. Inside the hull are floors called decks, and walls called bulkheads. These divide the hull into watertight compartments that can be closed off if the hull is damaged in an accident. That way the water cannot get into the ship.

Engines and Propellers

Most ships are powered by huge diesel engines inside the hull. The engine turns a driveshaft that spins a propeller under the stern (back) of the ship. Behind the propeller is the rudder, which turns left or right to steer the ship.

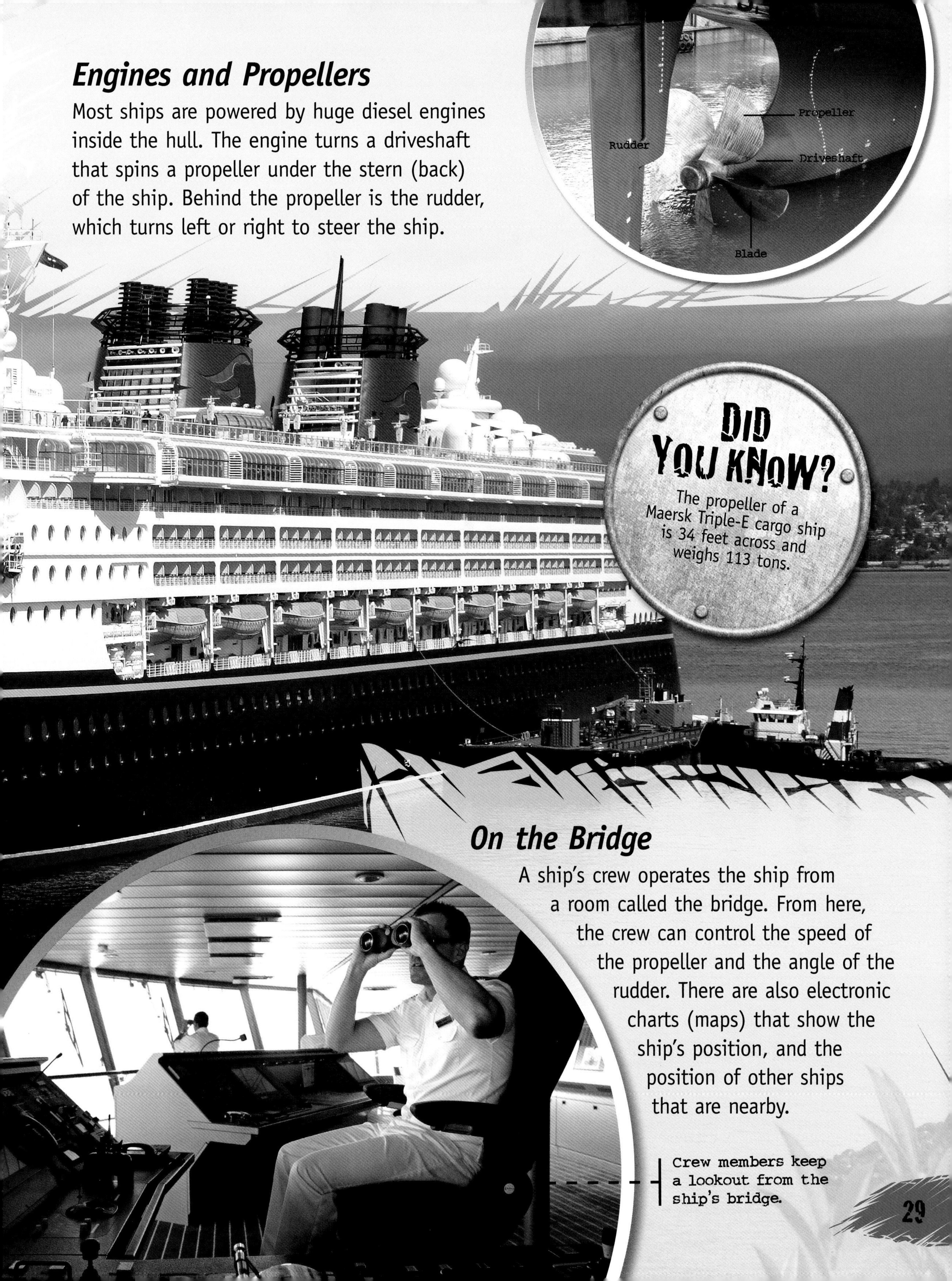

DID YOU KNOW?

The propeller of a Maersk Triple-E cargo ship is 34 feet across and weighs 113 tons.

On the Bridge

A ship's crew operates the ship from a room called the bridge. From here, the crew can control the speed of the propeller and the angle of the rudder. There are also electronic charts (maps) that show the ship's position, and the position of other ships that are nearby.

Crew members keep a lookout from the ship's bridge.

CARGO SHIPS

There are thousands of giant cargo ships crossing the world's seas and oceans, delivering cargo from one place to another. Many are general-purpose cargo ships, but others are specialized ships, such as container ships, oil tankers, and ferries.

Container ships normally have cranes onboard for loading, lifting, and moving the containers.

Container Movers

A container is a large metal box that can be carried by trucks, trains, and vast container ships. On ships, containers are carried in racks in the hull, and stacked on deck. They are loaded and unloaded at special container ports around the world.

Oil Tankers

The hull of an oil tanker is divided into many tanks filled with oil. There are also ballast tanks along each side of the ship that are filled with water to keep the ship balanced. Modern tankers have an inner hull and an outer hull (called a double hull), so that if the outer hull is damaged, oil cannot leak out.

A VLCC oil tanker (Very Large Crude Carrier).

Roll-on Roll-off

Ships and ferries that transport cars and vehicles are normally roll-on roll-off (RORO) vessels. There is a loading door at the bow (front) and stern (back) so that vehicles can drive on for loading and drive off for unloading.

This roll-on roll-off ferry has a bow door that opens for cars to drive on.

DID YOU KNOW?

The world's largest container ships, which carry nearly 20,000 containers, take 3.5 miles to slow to a complete stop from their top speed.

SAILING BOATS

Sailing boats and ships use the wind in their sails to push or propel them along. Sail power was the only form of boat power (except for rowing) for thousands of years, until steam engines were invented. Sailing boats are now used for leisure, and in some places as fishing boats.

DID YOU KNOW?

In 2013 the hydrofoil yacht Hydroptere reached a world-record speed for a sailing boat of nearly 61 miles per hour.

Parts of a Yacht

A sailing yacht is a boat that has a sail and is used for leisure or racing. It has a smooth hull so it can easily slip through the water. Under the hull is a heavy keel like a fin that prevents the boat tipping over too far to one side in strong winds. The sails are supported by masts and a boom. The boat is steered with a wheel or tiller (handle) attached to the rudder.

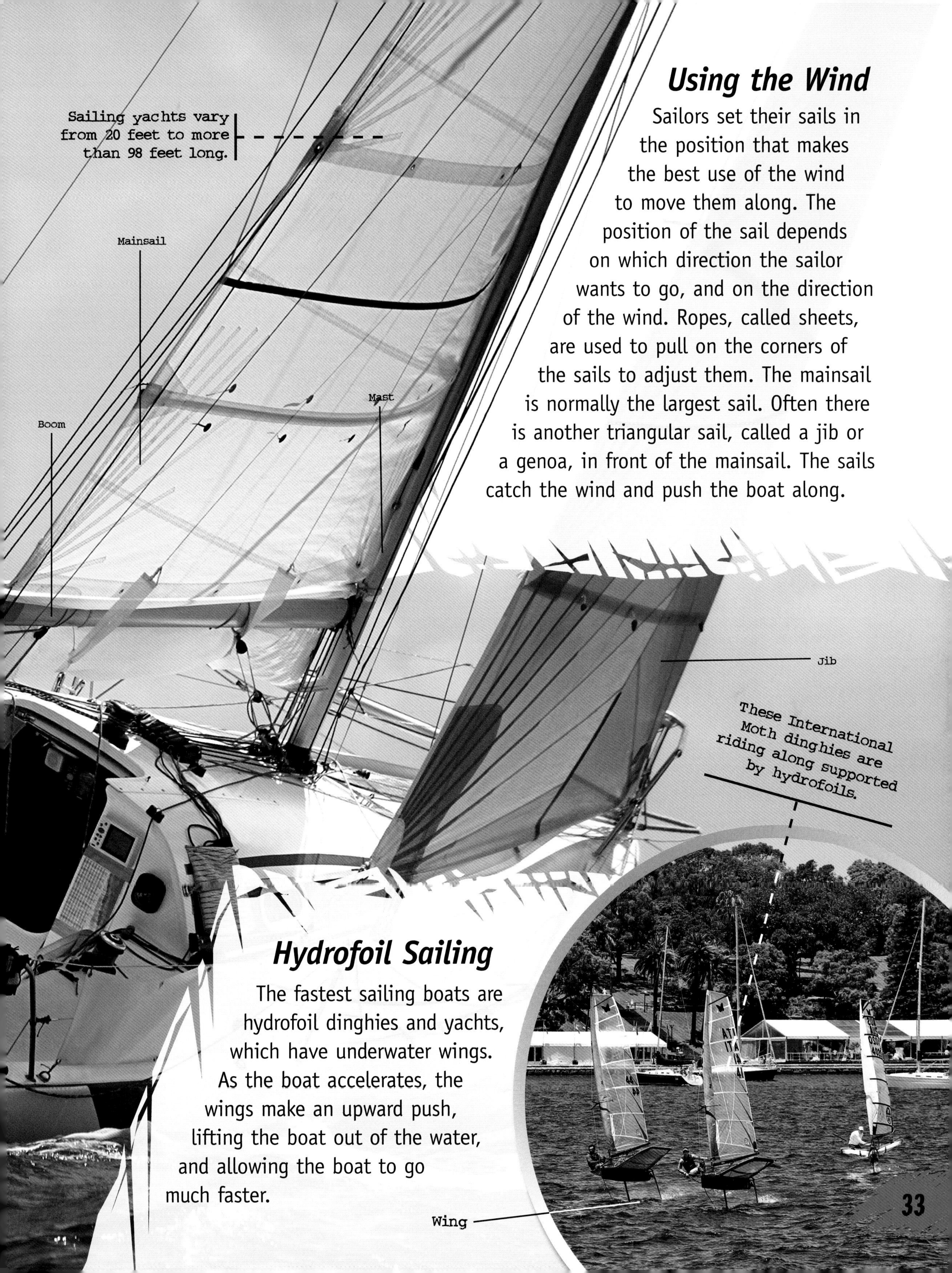
Sailing yachts vary
from 20 feet to more
than 98 feet long.
Mainsail
Boom
Mast
Jib
These International
Moth dinghies are
riding along supported
by hydrofoils.
Wing

Using the Wind

Sailors set their sails in the position that makes the best use of the wind to move them along. The position of the sail depends on which direction the sailor wants to go, and on the direction of the wind. Ropes, called sheets, are used to pull on the corners of the sails to adjust them. The mainsail is normally the largest sail. Often there is another triangular sail, called a jib or a genoa, in front of the mainsail. The sails catch the wind and push the boat along.

Hydrofoil Sailing

The fastest sailing boats are hydrofoil dinghies and yachts, which have underwater wings. As the boat accelerates, the wings make an upward push, lifting the boat out of the water, and allowing the boat to go much faster.

SPEEDBOATS

This racing speedboat is planing as it reaches its maximum speed.

Speedboats (also known as powerboats) are built to go super fast. Speedboats are used for fun, for racing, and as rescue boats. Their high speed is used to reach people in danger in the water as quickly as possible.

DID YOU KNOW?
The outboard engine of a Formula 1 speedboat is five times more powerful than the engine in a family car.

Water-jet Power

Some speedboats have water-jet power instead of propellers. The engine drives a pump, the pump sucks in water from under the craft, and then fires it out of the back in a jet. The jet of water pushes the boat forward.

Jet skis use water-jet power.

Engines at the Back

Most speedboats have an outboard motor, fixed to the stern (back) of the boat, with the propeller reaching down into the water. The engine powers the boat, and is also turned from side to side to steer the boat.

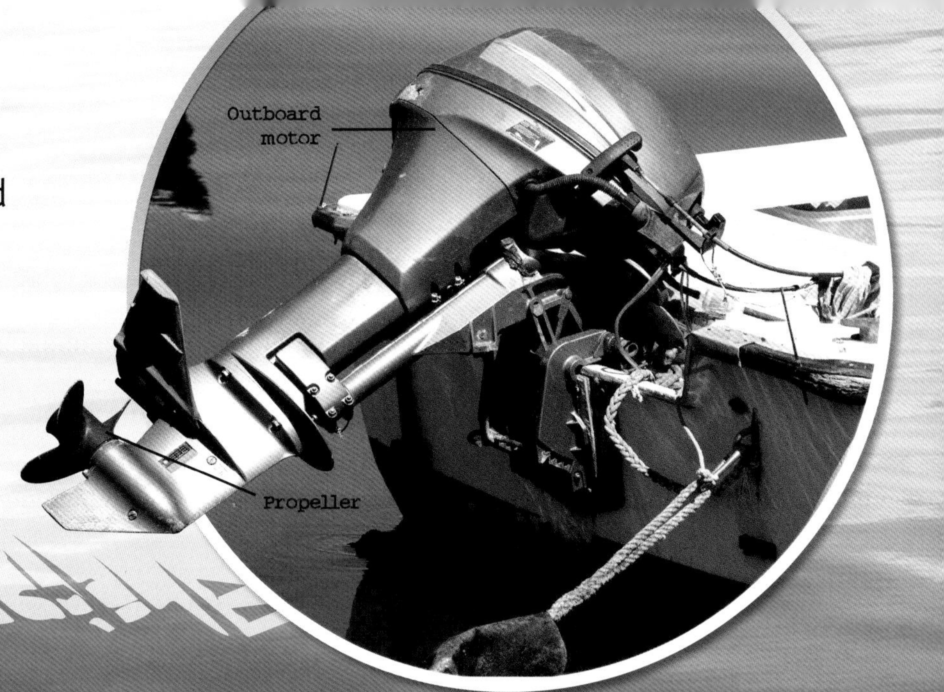

Skimming on the Water

As a speedboat picks up speed, the front of its hull rises out of the water and it skims on its flat bottom. This is called planing. With the hull mostly out of the water, the boat can reach very high speeds.

Racing Speedboats

A racing speedboat has a lightweight hull with two planing hulls side by side. At high speeds, air rushes between the hulls, and this helps to lift the boat higher out of the water. Only a tiny part of the hull touches the water as the boat skips across the waves.

Racing speedboats skimming across the water at top speed.

CANALS

A canal is a man-made waterway that goes across land. Some canals link one city to another city inland, some link cities to the sea, and others link one sea to another sea to make shortcuts for ships.

The Panama Canal is 48 miles long and has three sets of double locks.

Locks

A lock is a section of canal with gates at each end. The gates are opened to let boats into the lock. The level of water in the lock is changed to move the boat up or down so it can pass from one part of the canal to the other.

This is how a ship moves up through a lock to get to a higher level of a canal.

Panama Canal

The Panama Canal cuts through the country of Panama in Central America, linking the Atlantic Ocean and the Pacific Ocean. The huge locks move ships up and down between the different levels of the canal. New locks being built on the canal are 180 feet wide. Many cargo ships, called Panamax ships, are specially designed to fit through the canal.

DID YOU KNOW?

The 363-mile-long Erie Canal was one of America's greatest engineering feats. It opened in 1825, linking New York State to the Great Lakes.

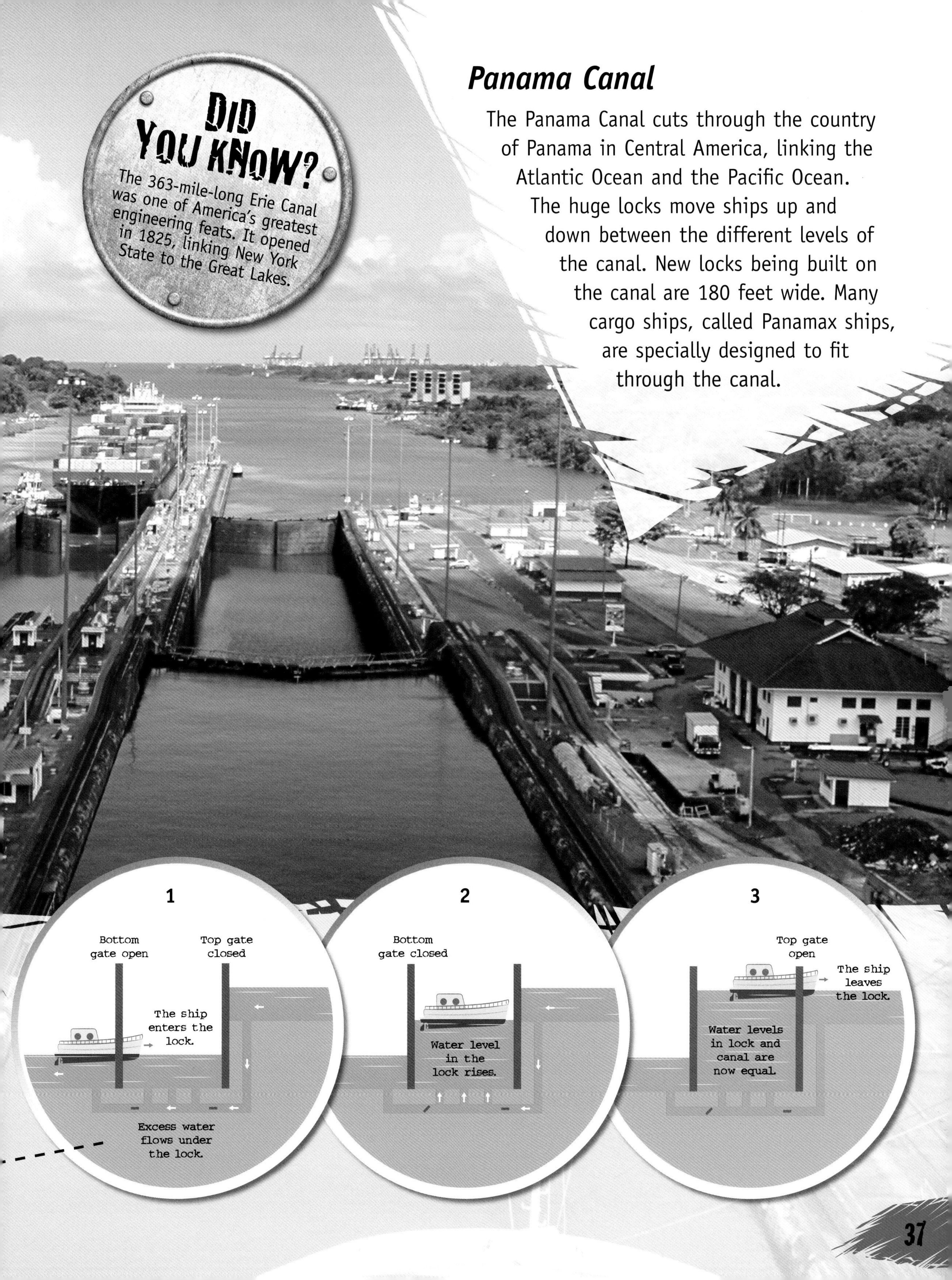
1
Bottom gate open
Top gate closed
The ship enters the lock.
Excess water flows under the lock.
2
Bottom gate closed
Water level in the lock rises.
3
Top gate open
The ship leaves the lock.
Water levels in lock and canal are now equal.

SUBMARINES AND SUBMERSIBLES

Submarines and submersibles are underwater crafts. A submarine is a large, manned vessel that can travel underwater. A submersible is a small underwater craft, with a small crew or no crew at all.

This is a diesel-electric submarine.

Submarine Parts

A diesel-electric submarine has a diesel engine, an electricity generator, batteries, and electric motors that drive the propeller. On the surface of the water, the engine drives the generator to make electricity for the motors and to recharge the batteries. When the submarine is submerged under the water, the motor runs from the batteries. A nuclear submarine has a nuclear reactor that heats water to produce steam, and the steam powers a **turbine** that turns the propeller.

DID YOU KNOW?

In 2012, film director James Cameron made a record-breaking dive 7 miles down, in his Deepsea Challenger submersible.

Diving and Surfacing

Submarines have ballast tanks in the hull. To dive under the water, the ballast tanks are flooded with water, making the submarine heavier. To return to the surface, air is pumped into the ballast tanks to push the water out, making the submarine lighter. Hydroplanes are like small wings that can be twisted to point up or down. They help the submarine to dive and surface, and are used to steer the submarine up and down as it moves underwater.

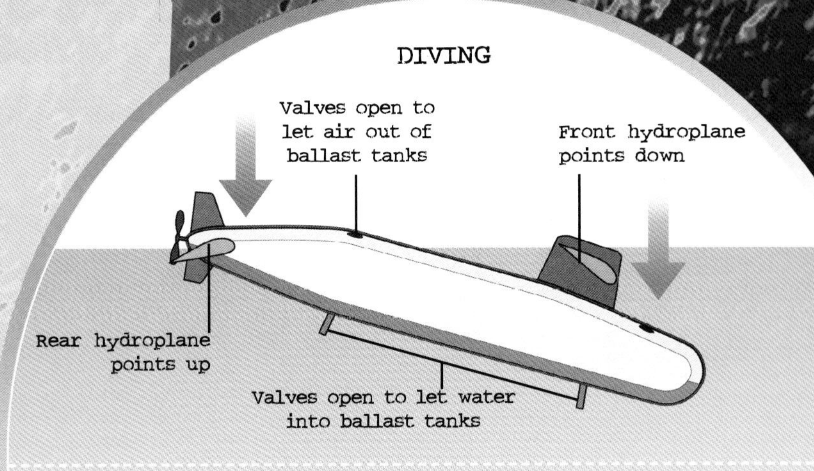

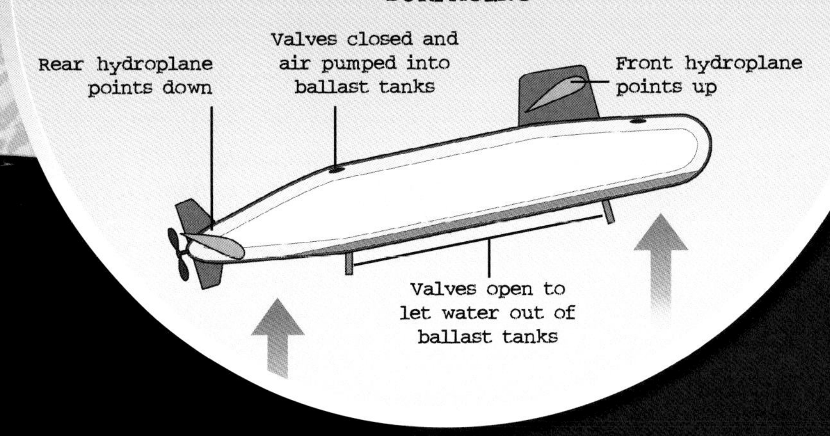

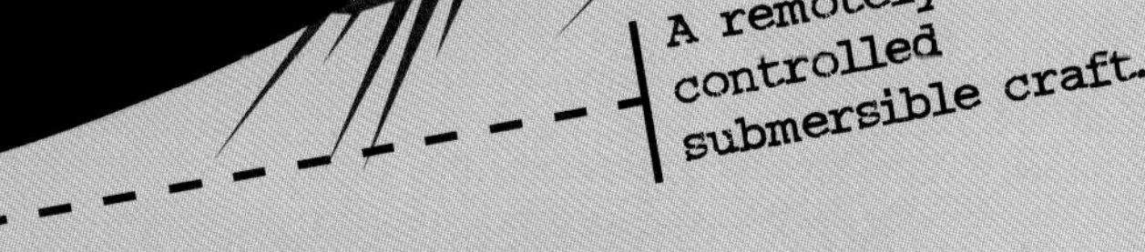
A remotely controlled submersible craft.

Submersible ROVs

Most submersibles are unmanned, remotely operated vehicles (ROVs). They are often used to support off-shore platforms (see page 136-137), the military, scientists studying the ocean, and filmmakers. Most ROVs are equipped with a camera, lights, and other tools such as a manipulator arm. They are controlled from a ship on the surface of the water and linked to the ship by a **tether** or cable. An ROV has small, powerful propellers called **thrusters** to make it dive and surface, and to drive it forward.

Thruster
Camera
Manipulator arm

BASICS OF FLIGHT

DID YOU KNOW?
Wings have flaps that come out to make the wing bigger as a plane flies slowly during take-off and landing.

An airplane is a flying machine with wings that keep it up in the air. Planes come in dozens of different shapes and sizes, but they all have the same basic parts.

Plane Parts

The body of a plane is called its fuselage. It is a framework covered with a thin metal skin. Wings are attached to the fuselage, one on each side. The tailplane is made up of a vertical fin and horizontal stabilizers. It keeps the plane flying straight and level. An engine pulls the plane through the air.

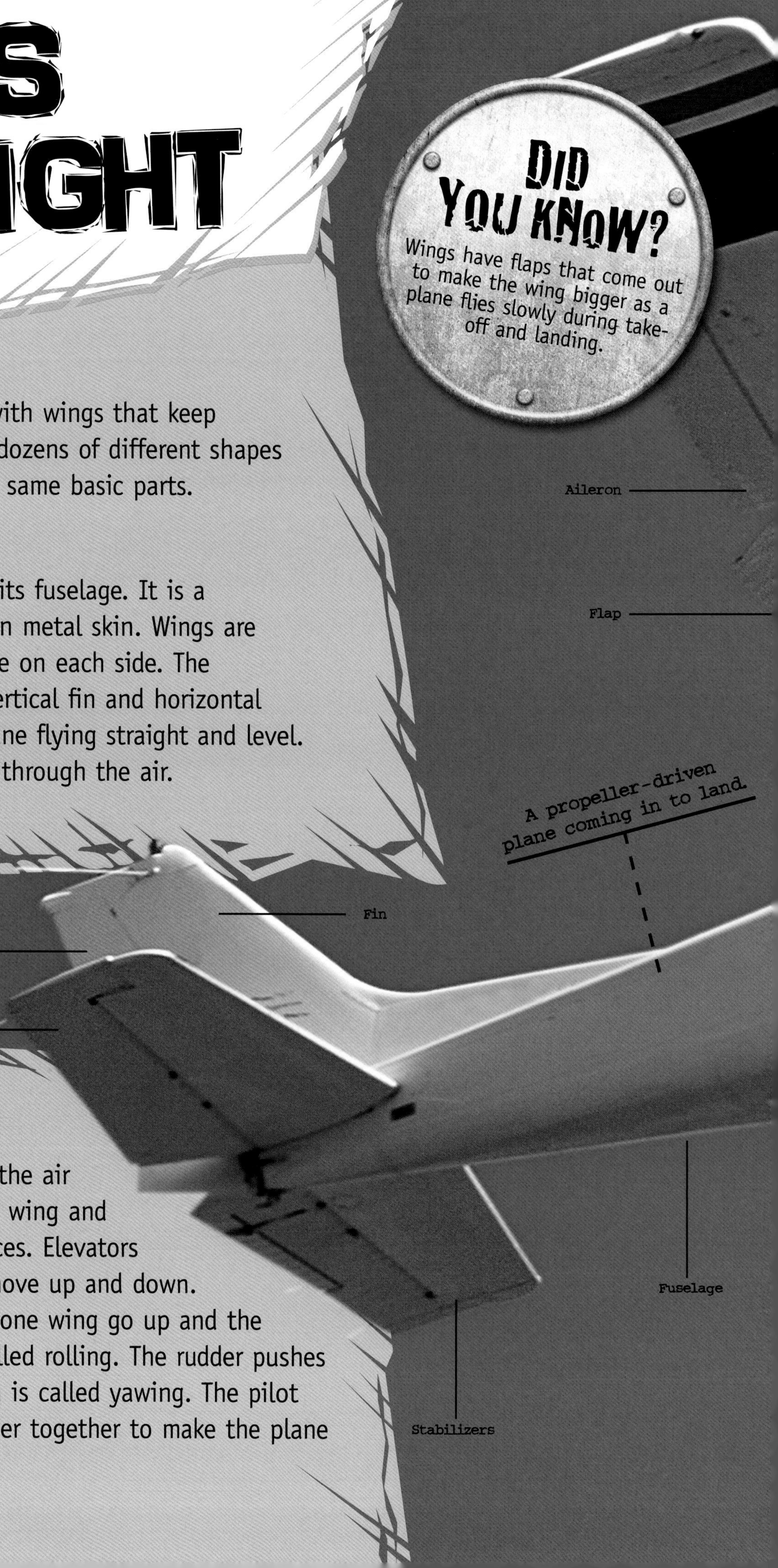

Aileron
Flap
Fin
Rudder
Elevator
Fuselage
Stabilizers

A propeller-driven plane coming in to land.

Flight Control

A pilot steers a plane through the air by moving hinged parts of the wing and tailplane, called control surfaces. Elevators on the tail make the plane move up and down. Ailerons on the wings make one wing go up and the other go down, which is called rolling. The rudder pushes the tail left or right, which is called yawing. The pilot uses the ailerons and rudder together to make the plane turn left or right.

In the Cockpit

A pilot's main controls are a control wheel or stick, and pedals. Pushing the wheel back and forth operates the elevators. Turning the wheel from side to side operates the ailerons. Pressing the pedals operates the rudder.

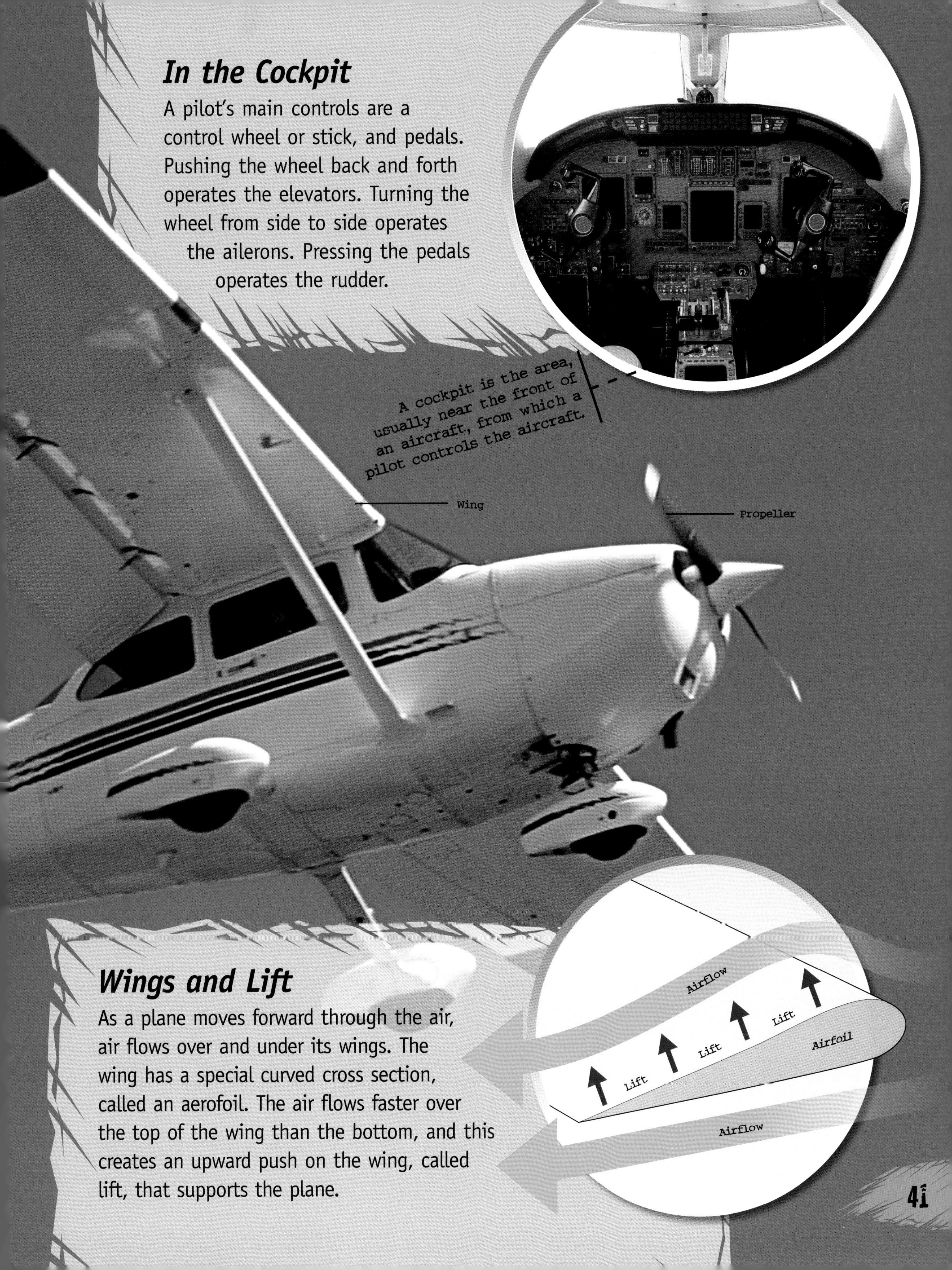
A cockpit is the area, usually near the front of an aircraft, from which a pilot controls the aircraft.
Wing
Propeller
Airflow
Lift
Lift
Lift
Airfoil
Airflow

Wings and Lift

As a plane moves forward through the air, air flows over and under its wings. The wing has a special curved cross section, called an aerofoil. The air flows faster over the top of the wing than the bottom, and this creates an upward push on the wing, called lift, that supports the plane.

BALLOONS AND AIRSHIPS

Planes and helicopters have wings or rotors that provide the lift needed to keep them in the air. They are known as heavier-than-air aircraft. Balloons and airships are lighter-than-air aircraft. They stay up because they float in the air.

Hot-air balloon flights and festivals are popular in Albuquerque, New Mexico.

A balloonist uses a burner to inflate a hot-air balloon.

Filling with Hot Air

The fabric bag that is filled with hot air is called an envelope. The basket connected to the balloon lies on its side when the envelope is empty. The hot air from a gas burner is blown into the envelope and as the envelope fills with hot air it slowly blows up like a balloon and the basket lifts off the ground into an upright position. Hot air is lighter than cool air so the envelope of hot air floats up through the cool air around it.

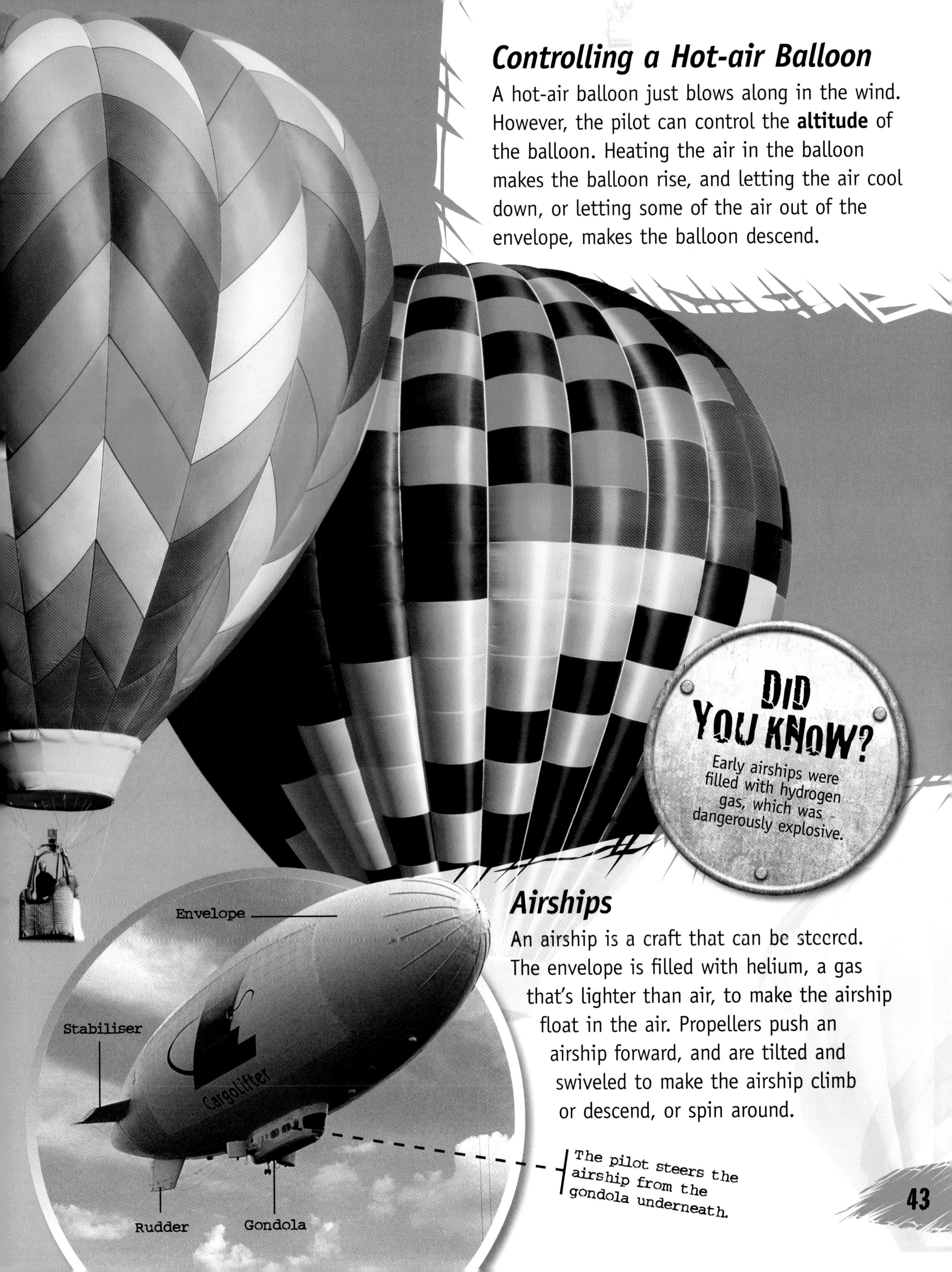

Controlling a Hot-air Balloon

A hot-air balloon just blows along in the wind. However, the pilot can control the **altitude** of the balloon. Heating the air in the balloon makes the balloon rise, and letting the air cool down, or letting some of the air out of the envelope, makes the balloon descend.

DID YOU KNOW?

Early airships were filled with hydrogen gas, which was dangerously explosive.

Airships

An airship is a craft that can be steered. The envelope is filled with helium, a gas that's lighter than air, to make the airship float in the air. Propellers push an airship forward, and are tilted and swiveled to make the airship climb or descend, or spin around.

The pilot steers the airship from the gondola underneath.

JET ENGINES

Small, slow planes, known as light aircraft, normally have internal combustion engines that work the propellers. Larger or faster planes, such as airliners and military jets, have powerful jet engines. A jet engine produces a stream of hot gases and air that rush out of the engine exhaust. The jet of hot gas rushing backwards produces thrust, which pushes the engine (and the plane) forward.

Turbofan Engines

The turbofan is the most common type of jet engine. At the front of the engine is a large fan that sucks in air. Air in the central part of the engine is squeezed by a compressor and then goes into a **combustion** chamber to feed the burning fuel. The burning fuel creates a stream of hot gases that rush through a turbine, and stream out of the engine's exhaust, creating thrust to propel the plane forward. The turbine turns the fan and compressor at the front of the engine.

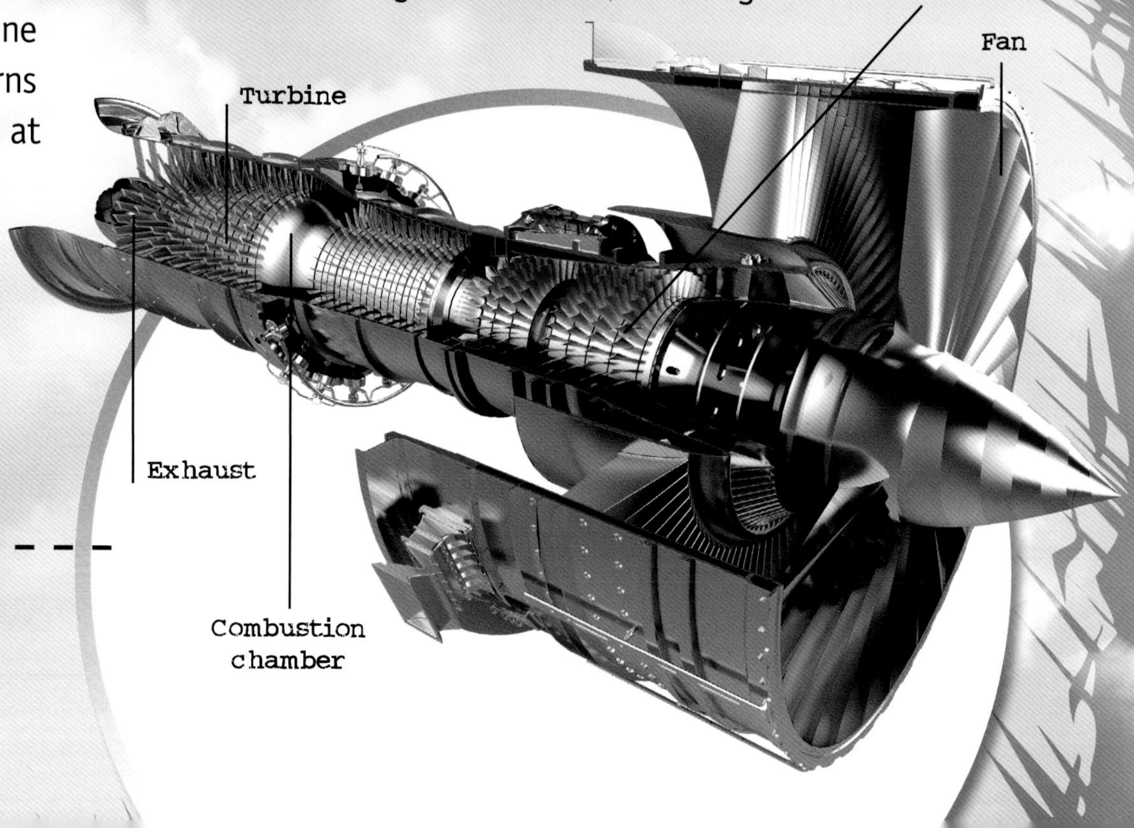

A cutaway picture of a turbofan engine shows the parts inside.

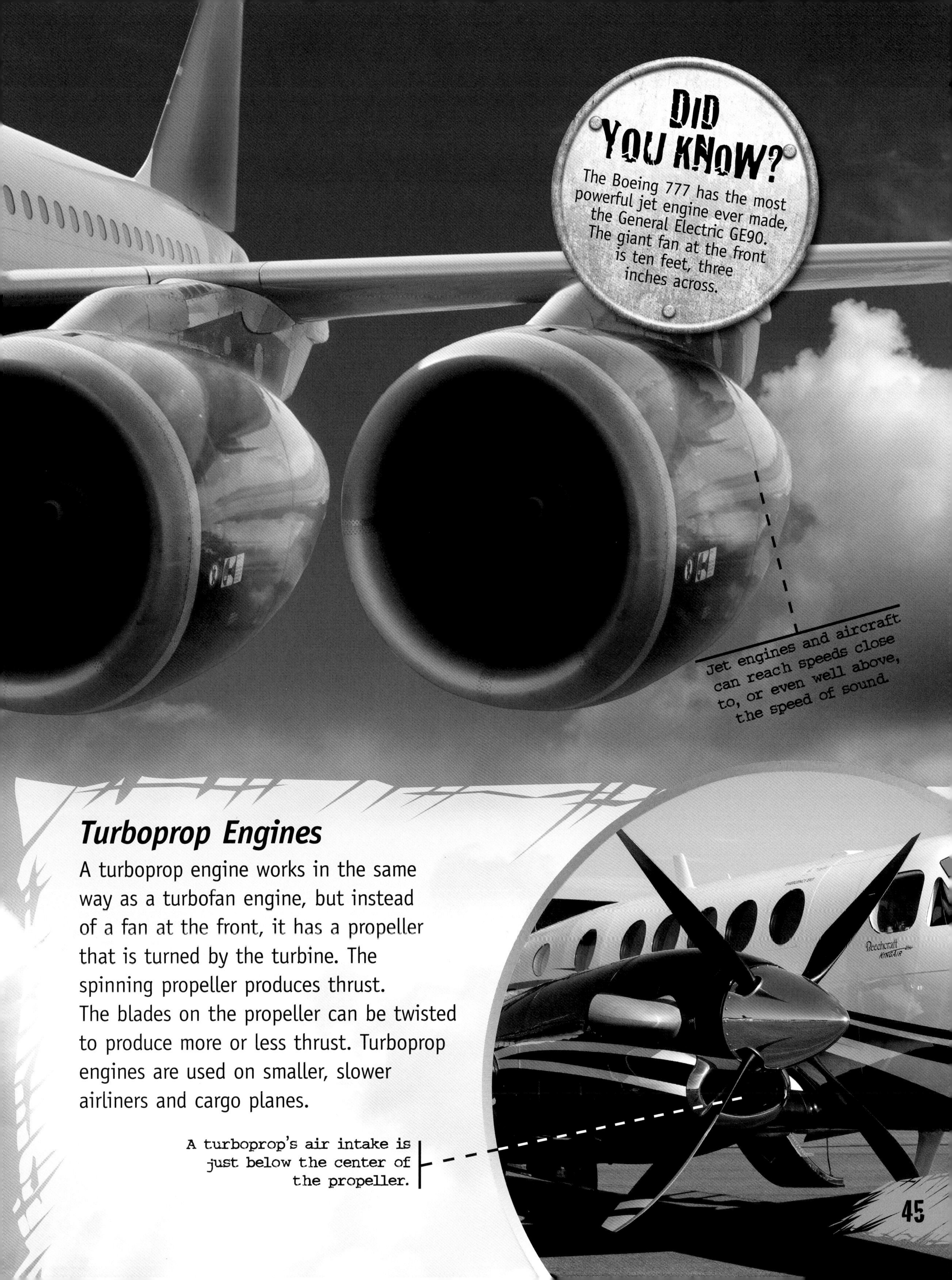

DID YOU KNOW?
The Boeing 777 has the most powerful jet engine ever made, the General Electric GE90. The giant fan at the front is ten feet, three inches across.
Jet engines and aircraft can reach speeds close to, or even well above, the speed of sound.

Turboprop Engines

A turboprop engine works in the same way as a turbofan engine, but instead of a fan at the front, it has a propeller that is turned by the turbine. The spinning propeller produces thrust. The blades on the propeller can be twisted to produce more or less thrust. Turboprop engines are used on smaller, slower airliners and cargo planes.

A turboprop's air intake is just below the center of the propeller.

AIRLINERS

Airliners are planes that carry passengers. Small airliners carry a few dozen passengers over short distances. Large airliners carry hundreds of passengers. These giant machines can fly almost halfway around the world without refueling.

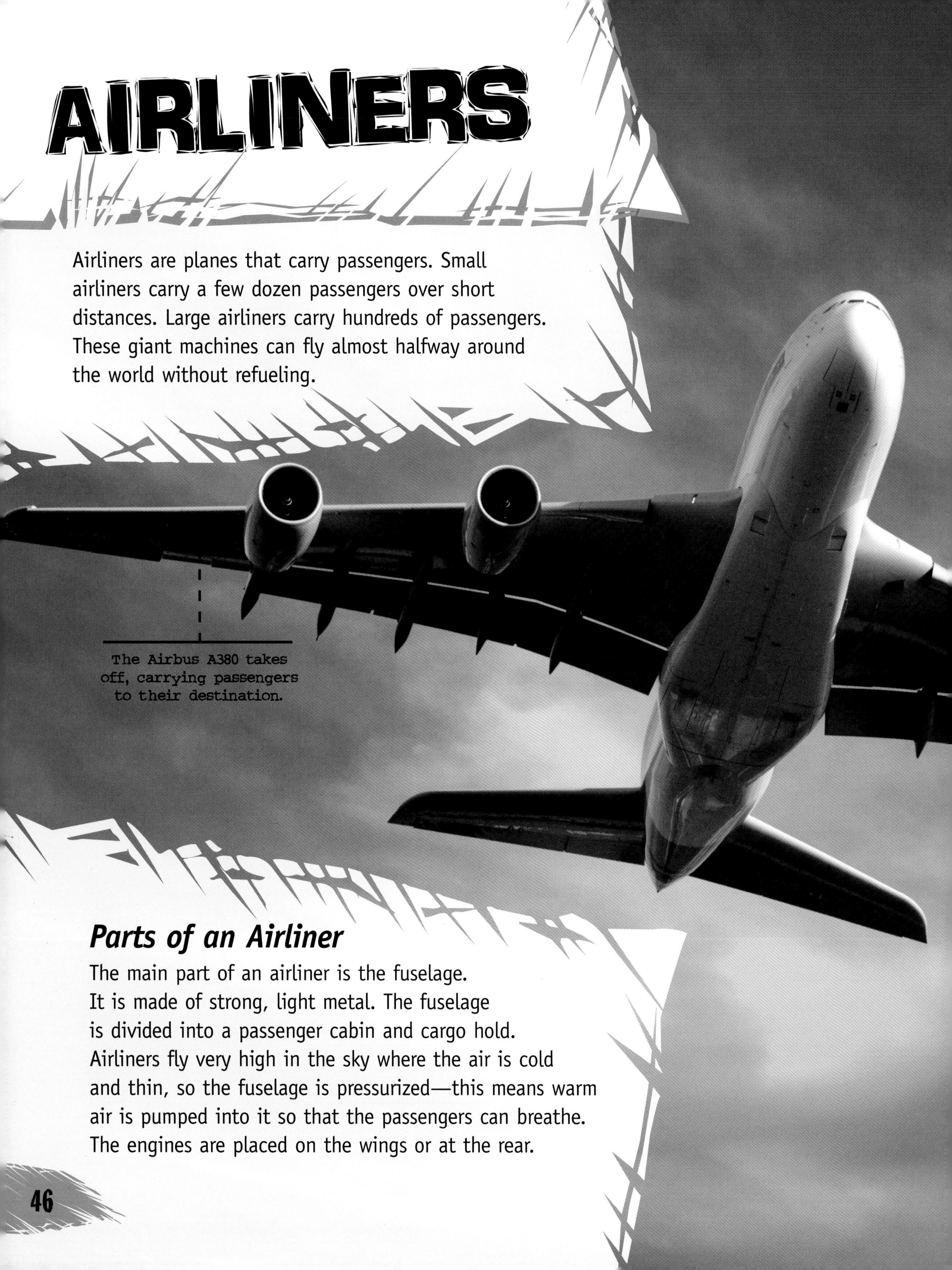

The Airbus A380 takes off, carrying passengers to their destination.

Parts of an Airliner

The main part of an airliner is the fuselage. It is made of strong, light metal. The fuselage is divided into a passenger cabin and cargo hold. Airliners fly very high in the sky where the air is cold and thin, so the fuselage is pressurized—this means warm air is pumped into it so that the passengers can breathe. The engines are placed on the wings or at the rear.

Flaps come out from
the rear of the wing
before landing.

Landing Approach

Airliners and other large planes slow down as they prepare to land. The pilot extends the flaps on the rear of the wing. This helps the wings create enough lift to keep the plane in the air while it slows down and prepares to land. An automatic landing system lands the plane.

DID YOU KNOW?

The Airbus A380 has a fuel capacity of 85,472 US gallons—enough for a family car to drive around the world about 120 times.

Safety Onboard

In an emergency, an airliner's exit doors contain slides that inflate or blow up when the doors are opened. Inflatable life jackets that contain a whistle to attract attention and a light for the dark are provided. The cabin also contains emergency oxygen masks that drop down from overhead lockers in case air escapes from the cabin.

Safety slides allow passengers to get off the plane quickly.

FAST JETS

Fighting aircraft fly much faster than airliners and cargo planes. They are designed to attack enemy targets. All have jet engines, and many can fly at **supersonic** speeds.

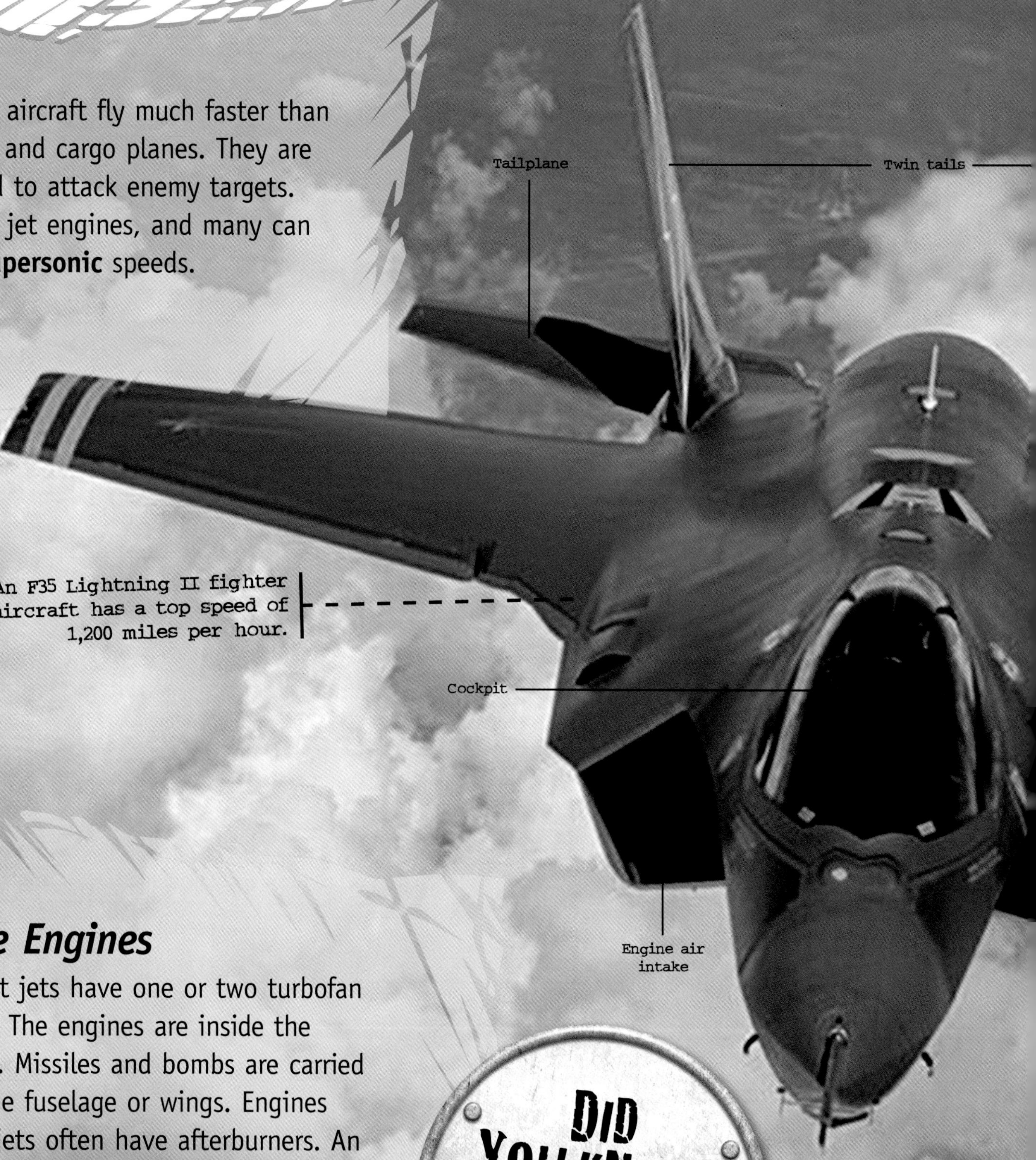

Tailplane

Twin tails

An F35 Lightning II fighter aircraft has a top speed of 1,200 miles per hour.

Cockpit

Engine air intake

Inside Engines

Most fast jets have one or two turbofan engines. The engines are inside the fuselage. Missiles and bombs are carried under the fuselage or wings. Engines on fast jets often have afterburners. An afterburner squirts fuel into the hot gases streaming from the back of the engine. The burning fuel adds power. Pilots switch on afterburners for a burst of speed, for a quick takeoff, or to climb steeply.

DID YOU KNOW?

Modern fighter jets would be impossible for pilots to control without the help of a computer.

Pilot Controls

Computers help the pilot to fly a fast jet. The pilot tells the flight computer when to climb, descend, or turn, and the computer moves the plane's controls to make the maneuver. The pilot tells the aircraft's navigation system where to go and a computer moves the plane's controls to take the plane there. Yet another computer tracks enemy targets using **radar**, and aims weapons ready to be fired.

The cockpit controls in an F35 Lightning II.

An AV 8B Harrier taking off vertically.

Vertical Takeoff

Some planes can take off and land vertically. To takeoff, the plane's engine exhaust nozzles turn to face downward, so the engine's thrust pushes the plane upward. After takeoff, the nozzles turn to face backward, and the plane moves into forward flight.

DRONES

A drone is a small, pilotless aircraft. Drones are also known as unmanned aerial vehicles (AUVs). Some are remote-controlled and some can fly themselves automatically.

Rotor

DID YOU KNOW?
Scientists have built and flown experimental micro drones the size of bees.

Quadcopter Drones

A quadcopter is a drone with four spinning rotors that lift it into the air like a helicopter. Power for the rotors comes from onboard batteries. Security forces, search-and-rescue teams, and filmmakers use quadcopters as 'eyes in the sky' for taking videos. Video is sometimes stored onboard, and sometimes sent to the ground by radio. Quadcopters are normally flown by an operator on the nearby ground.

Video camera

An MQ-9 Reaper military drone.

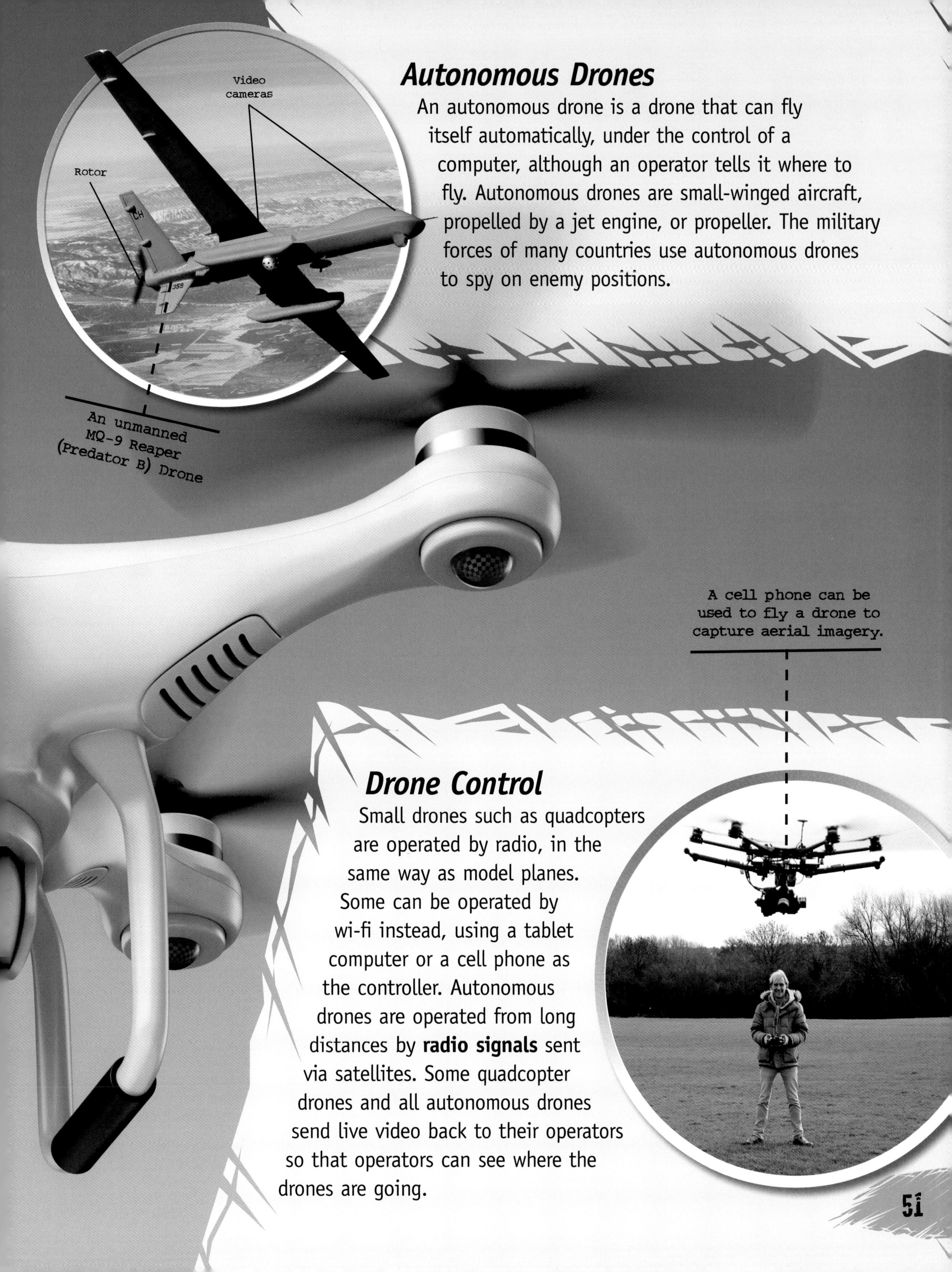

Video cameras
Rotor
An unmanned MQ-9 Reaper (Predator B) Drone
A cell phone can be used to fly a drone to capture aerial imagery.

Autonomous Drones

An autonomous drone is a drone that can fly itself automatically, under the control of a computer, although an operator tells it where to fly. Autonomous drones are small-winged aircraft, propelled by a jet engine, or propeller. The military forces of many countries use autonomous drones to spy on enemy positions.

Drone Control

Small drones such as quadcopters are operated by radio, in the same way as model planes. Some can be operated by wi-fi instead, using a tablet computer or a cell phone as the controller. Autonomous drones are operated from long distances by **radio signals** sent via satellites. Some quadcopter drones and all autonomous drones send live video back to their operators so that operators can see where the drones are going.

HELICOPTERS

The controls of a helicopter.

A helicopter is an aircraft that is lifted into the air by a spinning rotor. Helicopters can take off and land in a small area, where other aircraft can't. They can also hover in the same spot in the air, so they are good for jobs such as filming, and rescuing people from the sea or mountains after accidents.

Rotor blade

Turboshaft engine

Fuselage

Helicopter Parts

A helicopter's main rotor is at the top of its fuselage. The rotor is made up of a central hub and two or more rotor blades. A turboshaft engine turns the rotor. The engine makes the main rotor spin, but this also pushes the fuselage in the opposite direction, which makes the helicopter spin. The rotor blades work like wings, so when the rotor is spinning, they lift the helicopter. The small tail rotor helps to keep the helicopter moving in a straight line so it doesn't spin.

Pedals
Cyclic stick
Collective lever
RAF
RESCUE
A twin-rotor Chinook helicopter.
This rescue helicopter hovers in the air as a person is lifted to safety from an accident at sea.

Pilot Controls

A pilot flies a helicopter with a lever called the collective, a stick called the cyclic, and two pedals. The collective controls how much lift the main rotor blades make. Moving it up or down makes the helicopter go up or down. The cyclic makes the helicopter fly forward, backward, or sideways. The pedals make the helicopter turn one way or the other.

DID YOU KNOW?

Helicopters are sometimes known as "choppers" because of the chop-chop noise made by their rotors.

Twin Rotors

Some large helicopters have two rotors, one at the back and one at the front. Two rotors provide plenty of lift to transport heavy loads. The rotors also spin in opposite directions, and this means a twin-rotor helicopter doesn't need a tail rotor.

AIR TRAFFIC CONTROL

There are thousands of airliners and other aircraft up in the sky at the same time. The aircraft are taking off, climbing, cruising, descending, and landing. An air traffic control system keeps all these aircraft a safe distance apart.

Control Towers

All airports have a control tower where air traffic controllers communicate with the pilots by radio to tell them when to take off and when to land. After planes take off, different controllers take over, telling pilots to climb or descend, speed up or slow down, or turn left or right.

Radar

The word "radar" stands for radio detection and ranging. A radar system detects aircraft in the sky and a screen shows where the aircraft are. The radar does this using an **aerial** that sends out **radio waves**, which detect waves that bounce back from the other aircraft.

A radar aerial on top of an airport control tower.

Coming in to Land

Large airports have an instrument landing system. This sends out radio waves that the pilot can follow using navigation equipment on the plane. This helps the pilot find the runway, even in the dark or fog. Many aircraft have an automatic landing system that can land the aircraft without the pilot's help.

Pilots in the plane's cockpit ready to land.

Aircraft are shown on a computer screen along with information about them.

DID YOU KNOW?

Air traffic controllers around the world handle around 100,000 airliner flights every day.

ROCKETS

A rocket is a missile, spacecraft, aircraft, or any other machine that is propelled by a rocket engine. Rockets don't need a supply of air as they have to work in space where there is no air.

DID YOU KNOW?

The largest Space Launch System rocket is 384 feet tall and weighs 3,250 tons at liftoff.

Parts of a Space Rocket

The largest parts of any rocket are the fuel tanks, or containers for the engine (or engines). In a two-stage rocket, each stage has its own engine. The first stage engine fires for the first few minutes of a flight, then stage one breaks away from the rest of the rocket. The stage two engine takes over until the rocket is in space, and then the stage two engine breaks away.

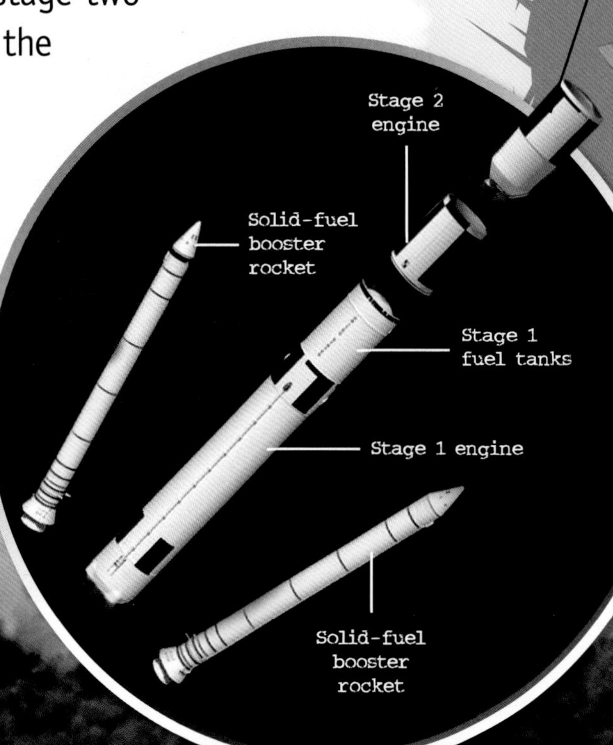

NASA's Space Launch System will launch cargo and astronauts into space. This is a cargo space rocket.

Ariane 5 takes off on July 1, 2009, in Kourou, French Guyana.

The five rocket engines of the first stage of a Saturn V rocket.

Liquid-fuel Rocket Engines

Rocket engines either have liquid fuel or solid fuel. There are two tanks of fuel in a liquid-fuel rocket. They are pumped into the combustion chamber where they produce hot gases. The gases rush out of the nozzle.

Solid-fuel Rocket Engines

In a solid-fuel rocket, the fuel is a powder. Once the powder is lit, it burns away until there is none left. A firework rocket has a simple solid-fuel motor that uses gunpowder for fuel. Booster rockets used on spacecraft are also solid-fuel rockets.

SPACECRAFT

In addition to taking astronauts into space and safely back to Earth, a spacecraft must keep astronauts warm, provide them with air to breathe, and protect them from dangerous **radiation**, such as gamma rays coming from the Sun.

DID YOU KNOW?

As a spacecraft re-enters Earth's atmosphere from space, it is traveling at around 17,500 miles per hour.

NASA's Orion is a new spacecraft being developed to take astronauts to the International Space Station, and eventually to Mars.

Spacecraft Parts

A spacecraft is launched into space by a large rocket. When it reaches space, the rocket separates and falls back to Earth. Spacecraft also have their own small rocket engines that are fired to speed them up or slow them down. A spacecraft contains all the equipment needed to keep the crew alive, such as oxygen supplies and water tanks.

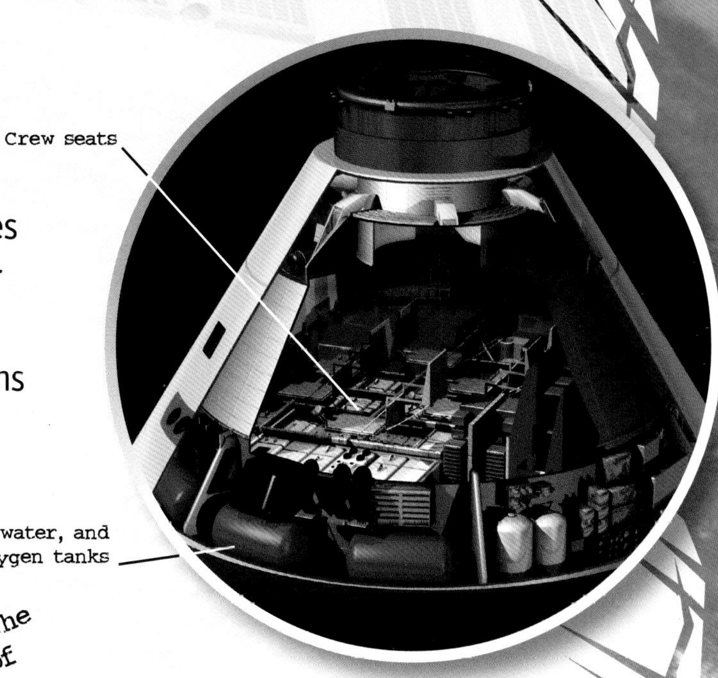

Inside the cabin of an Orion spacecraft.

Flight Controls

Astronauts have controls that they use to steer their craft through space and to work the spacecraft's systems. Computer screens give the astronauts information about how the craft is working. Most systems can be operated remotely by controllers on Earth, too.

Back to Earth

To return to Earth, a spacecraft fires its engines to slow down, which makes it drop out of space and re-enter the Earth's atmosphere. As it hurtles through the atmosphere, friction from the air makes the exterior red hot. A heat shield protects the spacecraft and crew from the heat. When it has slowed down enough, parachutes are released for a gentle landing.

Several large parachutes are needed to slow a heavy spacecraft.

INTERNATIONAL SPACE STATION

The International Space Station (ISS) is a space station that **orbits** the Earth. Astronauts live and work in the ISS, carrying out science experiments and learning about living in space.

The International Space Station is the size of a soccer field.

Getting There

Astronauts travel to the ISS and back to Earth again by spacecraft that park at docking points on the ISS. The docking points have hatches that the astronauts climb through to get into the station.

Parts of the ISS

The ISS is made up of connected sections called modules. Some modules contain living space, some are used for storage, and some are science laboratories. The ISS needs lots of electrical power for its lighting, communications equipment, and science laboratories. This power comes from huge panels of **solar cells** called photovoltaic arrays. As the ISS moves around Earth, its photovoltaic arrays turn automatically to face the Sun. They turn the sunlight into electricity.

DID YOU KNOW?
The International Space Station orbits 220 miles above Earth's surface, and completes one orbit every 90 minutes.
Photovoltaic array
The Canadarm2 on the ISS is used to move astronauts when they are on space walks.

Robotic Arms

The ISS has two robotic arms that support astronauts working in space. They also move equipment and supplies around the station and help with any maintenance or fixing. The base of one of the arms, the Canadarm2, moves on tracks so that the arm can reach almost all of the areas of the station.

SPACESUITS

When astronauts work outside their spacecraft making repairs, they wear special spacesuits to protect them from the dangers of space—the lack of air, the cold, and the deadly radiation.

Spacesuit Parts

A spacesuit is made up of a stiff upper torso, a lower torso, arms, gloves, and a helmet. These sections connect together with strong, airtight joints. The fabric has a fireproof layer, a waterproof layer, an insulating layer, and an armored layer that protects against tiny flying rocks called micrometeoroids. Even though it is extremely cold in space, an astronaut can get hot when working hard or from being exposed to the sun. Part of a spacesuit is a set of underwear that contains tubes of water to cool the skin.

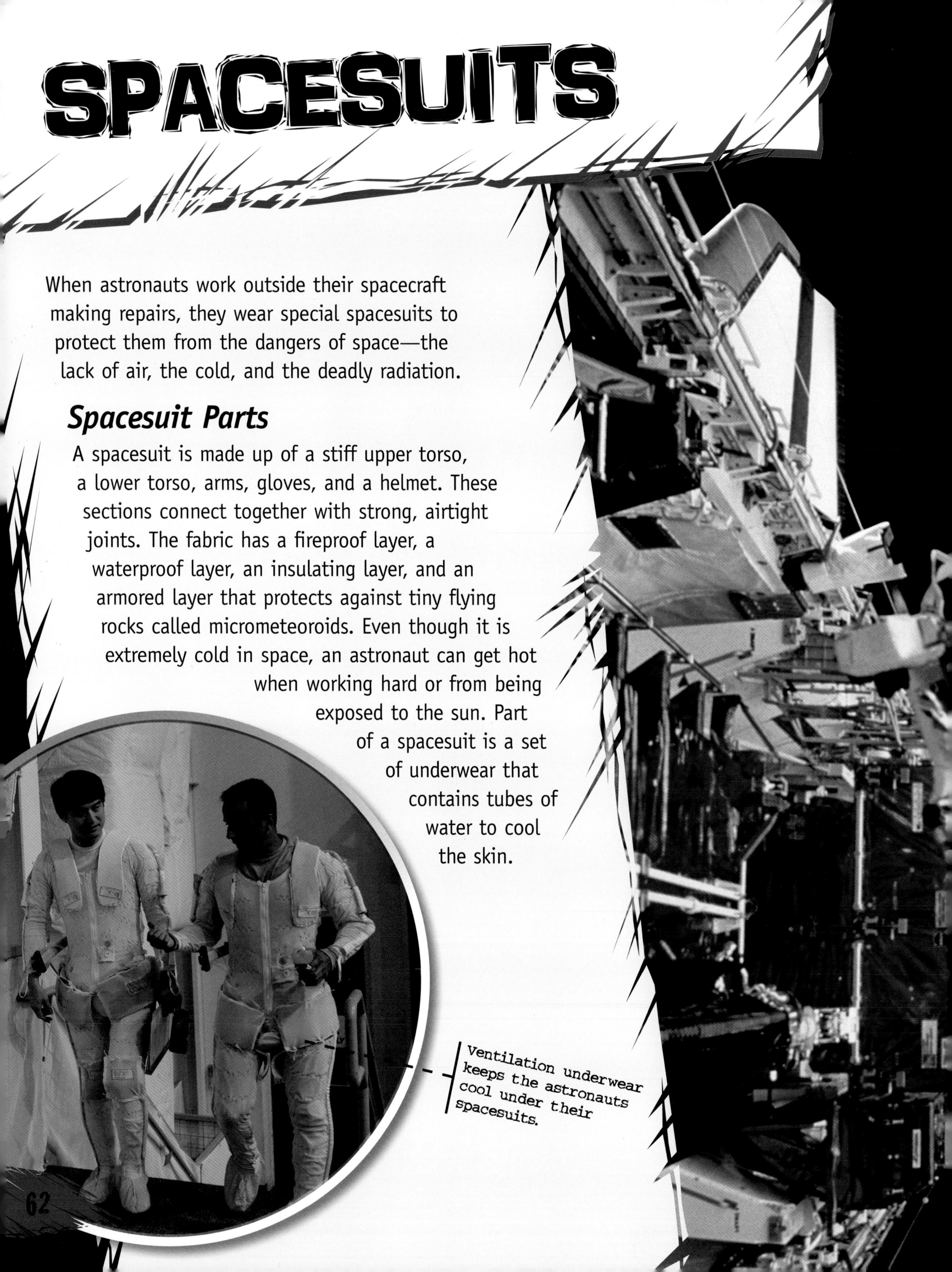

Ventilation underwear keeps the astronauts cool under their spacesuits.

Head Protection

A spacesuit helmet is supplied with oxygen for the astronaut to breathe from tanks in the backpack. The astronaut's face is covered with a thick, plastic visor.

Spacesuit Backpack

A spacesuit backpack contains oxygen tanks, batteries, work tools, water-cooling equipment for the underwear, and a radio. A small jetpack is often attached to the backpack. If the astronaut's tie breaks, he or she fires thrusters in the jetpack to get back to the spacecraft quickly.

An astronaut in a spacesuit on a spacewalk outside the International Space Station.

A spacesuit visor is covered in a thin layer of gold to protect the astronaut's eyes against the sun's harmful rays.

This spacesuit backpack is called a Primary Life Support Subsystem.

Spacesuit boots connect to a robotic arm.

DID YOU KNOW?
An astronaut's blood would boil if he or she went into space without a spacesuit.

SATELLITES

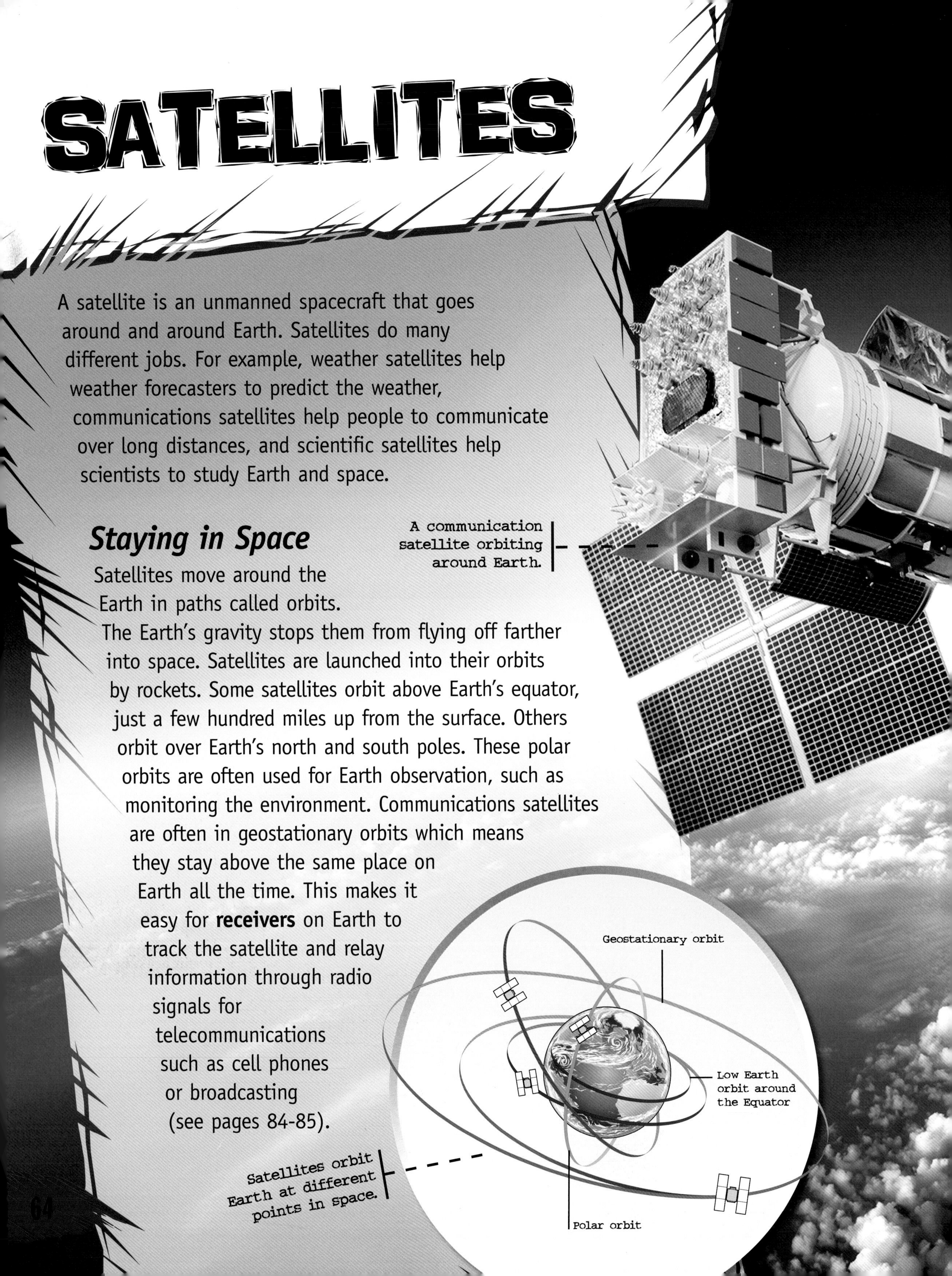

A satellite is an unmanned spacecraft that goes around and around Earth. Satellites do many different jobs. For example, weather satellites help weather forecasters to predict the weather, communications satellites help people to communicate over long distances, and scientific satellites help scientists to study Earth and space.

A communication satellite orbiting around Earth.

Staying in Space

Satellites move around the Earth in paths called orbits. The Earth's gravity stops them from flying off farther into space. Satellites are launched into their orbits by rockets. Some satellites orbit above Earth's equator, just a few hundred miles up from the surface. Others orbit over Earth's north and south poles. These polar orbits are often used for Earth observation, such as monitoring the environment. Communications satellites are often in geostationary orbits which means they stay above the same place on Earth all the time. This makes it easy for **receivers** on Earth to track the satellite and relay information through radio signals for telecommunications such as cell phones or broadcasting (see pages 84-85).

Satellites orbit Earth at different points in space.

Solar panels
EMC

DID YOU KNOW?
At any time, there are about 2,250 satellites in orbit around Earth.

Satellite Parts

The parts that a satellite needs depend on its job. For example, a weather satellite will have equipment to measure the temperature of Earth's surface and cameras to photograph clouds. Nearly all satellites have solar cells that turn sunlight into the electricity they need to work. All satellites have communications aerials to send and receive data from control stations on Earth.

Ground Stations

Satellites communicate with Earth using radio signals. Signals are beamed to a satellite from a large radio dish called a ground station. Dishes also collect radio signals sent down to Earth by satellites.

A satellite ground station.

GPS

The letters GPS are short for Global Positioning System. GPS is used for finding places on land and at sea, for tracking the position of moving objects, and for making maps.

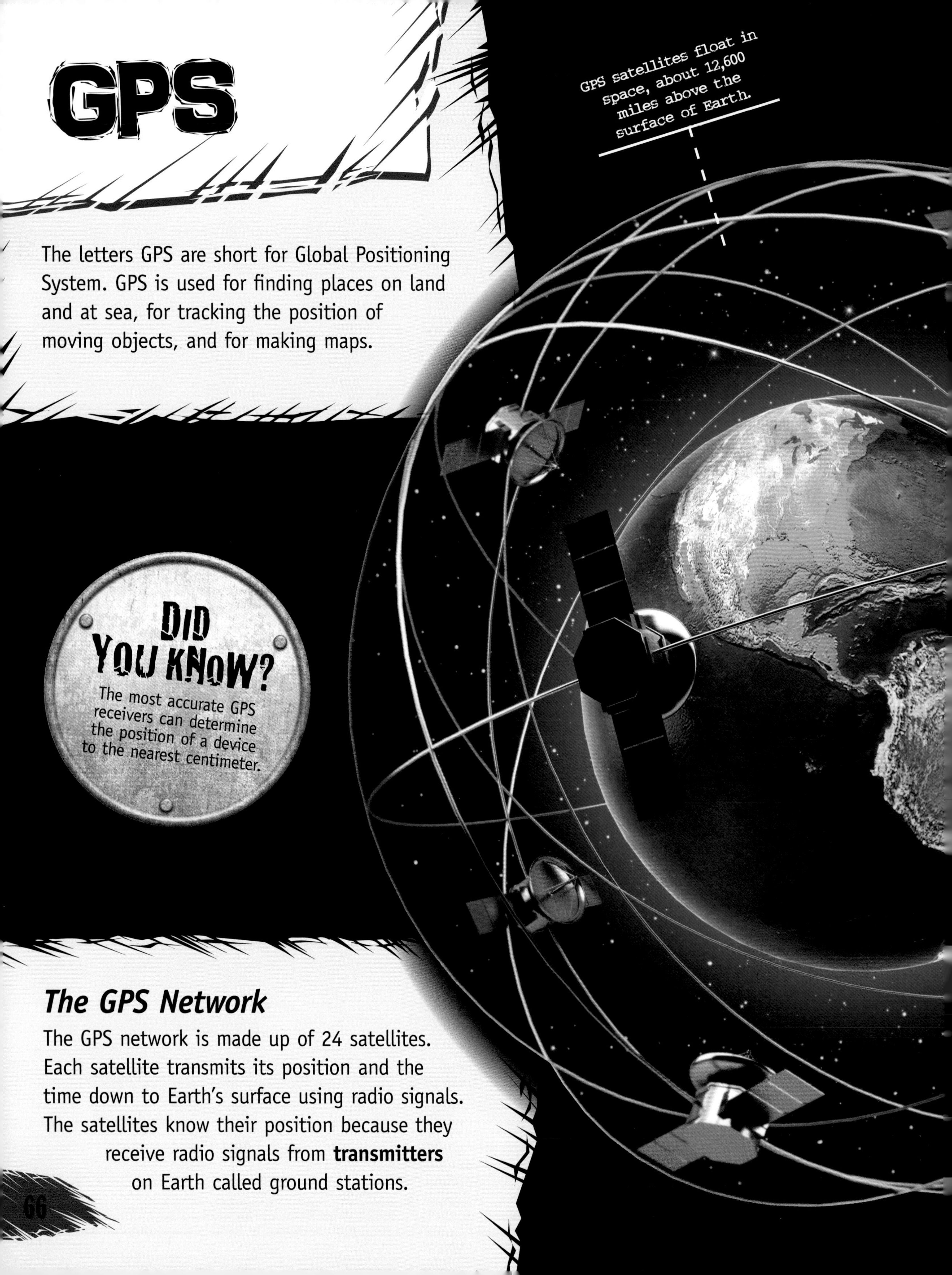

GPS satellites float in space, about 12,600 miles above the surface of Earth.

DID YOU KNOW?

The most accurate GPS receivers can determine the position of a device to the nearest centimeter.

The GPS Network

The GPS network is made up of 24 satellites. Each satellite transmits its position and the time down to Earth's surface using radio signals. The satellites know their position because they receive radio signals from **transmitters** on Earth called ground stations.

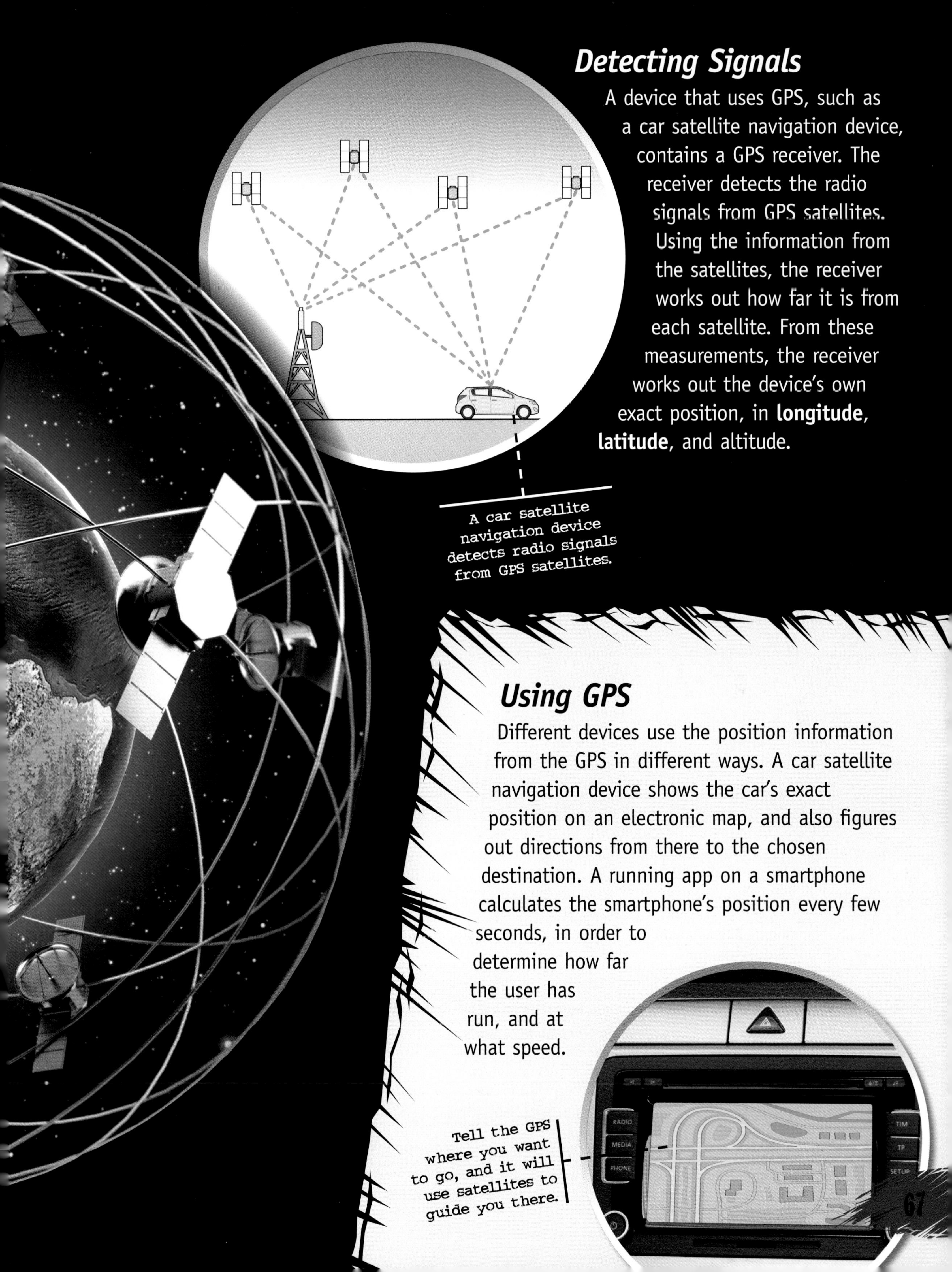

Detecting Signals

A device that uses GPS, such as a car satellite navigation device, contains a GPS receiver. The receiver detects the radio signals from GPS satellites. Using the information from the satellites, the receiver works out how far it is from each satellite. From these measurements, the receiver works out the device's own exact position, in **longitude**, **latitude**, and altitude.

A car satellite navigation device detects radio signals from GPS satellites.

Using GPS

Different devices use the position information from the GPS in different ways. A car satellite navigation device shows the car's exact position on an electronic map, and also figures out directions from there to the chosen destination. A running app on a smartphone calculates the smartphone's position every few seconds, in order to determine how far the user has run, and at what speed.

Tell the GPS where you want to go, and it will use satellites to guide you there.

SPACE PROBES

Space probes are unmanned spacecraft. Scientists send probes into space to explore other planets, moons, asteroids, and comets in our solar system. The probes send information and images back to Earth. Most of what we know about the solar system has come from probes.

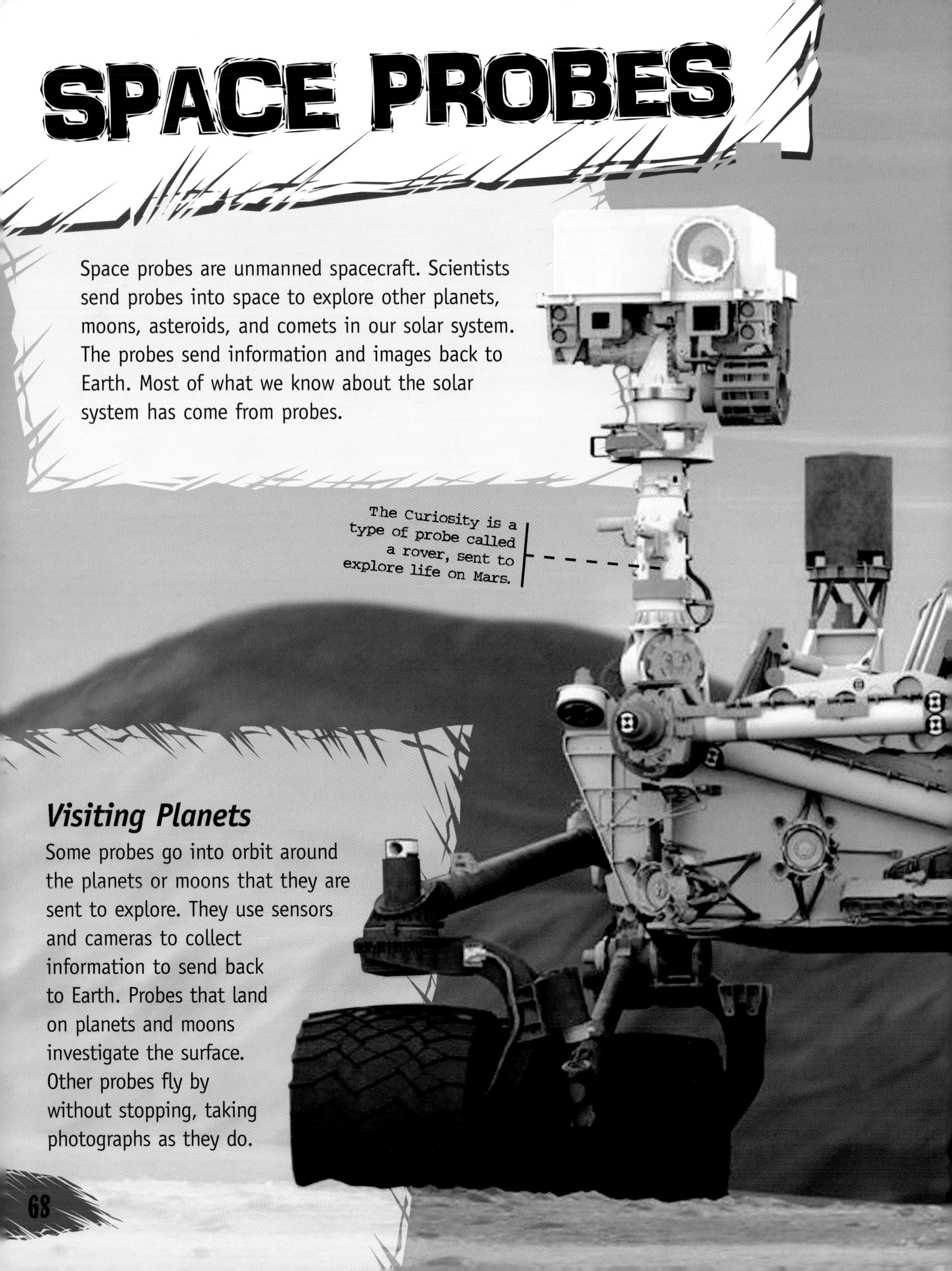

The Curiosity is a type of probe called a rover, sent to explore life on Mars.

Visiting Planets

Some probes go into orbit around the planets or moons that they are sent to explore. They use sensors and cameras to collect information to send back to Earth. Probes that land on planets and moons investigate the surface. Other probes fly by without stopping, taking photographs as they do.

Probe Power

Probes need electricity to work scientific instruments, cameras, moving parts such as robotic arms and wheels, and the antenna to transmit information back to Earth. On most probes, this electricity comes from solar cells, which turn sunlight into electricity. Probes that travel to the outer solar system, far from the sun, make electricity from tiny nuclear power plants onboard.

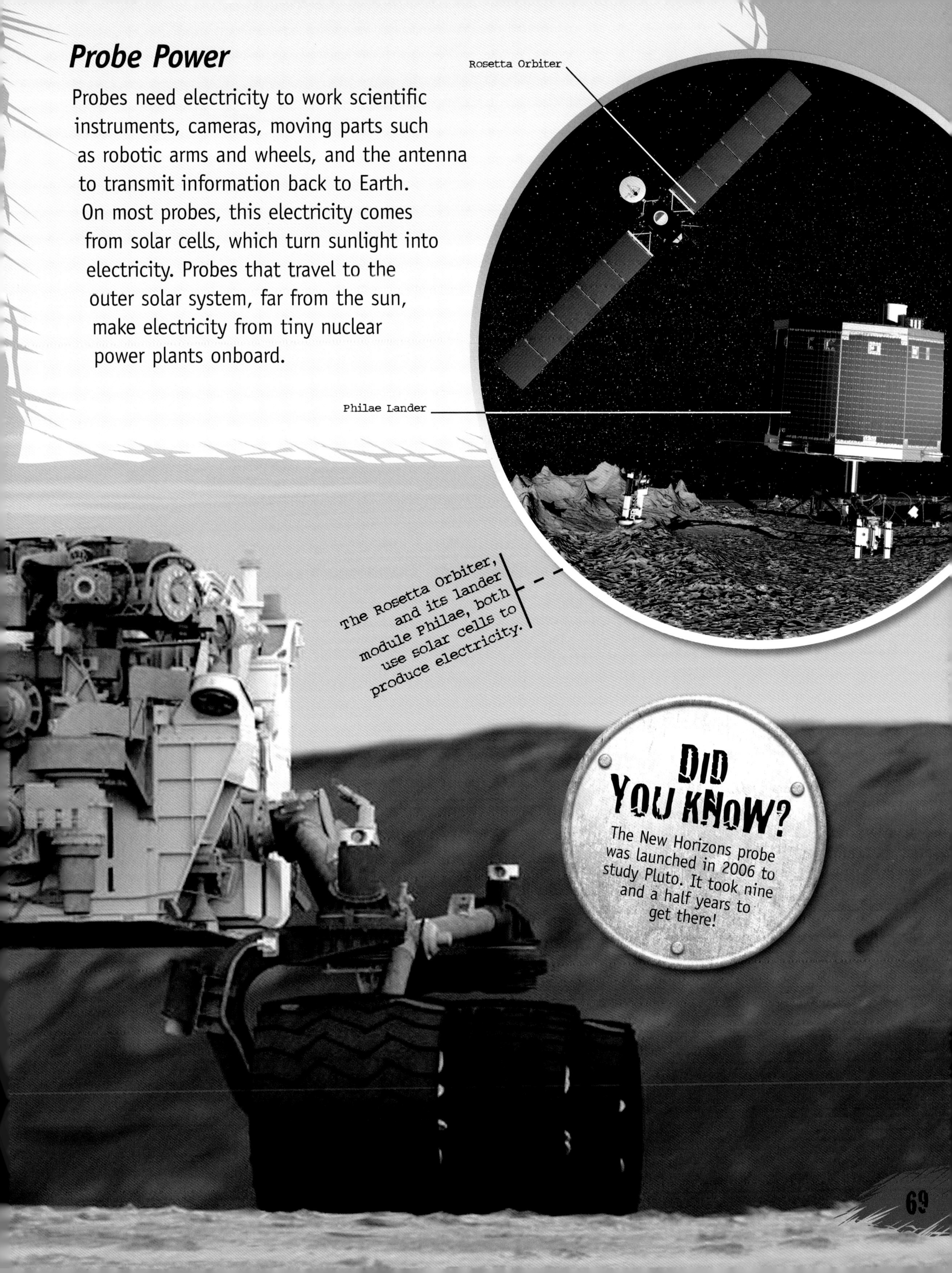

The Rosetta Orbiter, and its lander module Philae, both use solar cells to produce electricity.

DID YOU KNOW?

The New Horizons probe was launched in 2006 to study Pluto. It took nine and a half years to get there!

ELECTRONICS

Many devices we use every day, such as mobile phones, tablet computers, and digital watches, are electronic devices. These devices need electric circuits that control how electricity flows and how it operates each part of the device, such as the screen of a mobile phone.

Circuit Boards

Electric circuits are made up of different electronic components working together. Each component controls the electricity in a different way. The most common components are resistors, capacitors, transistors, and diodes. A resistor slows the flow of electricity; a capacitor stores electricity; a transistor turns electricity on and off like a switch, and a diode lets electricity flow in one direction only.

Components are connected by metal tracks on the circuit board.

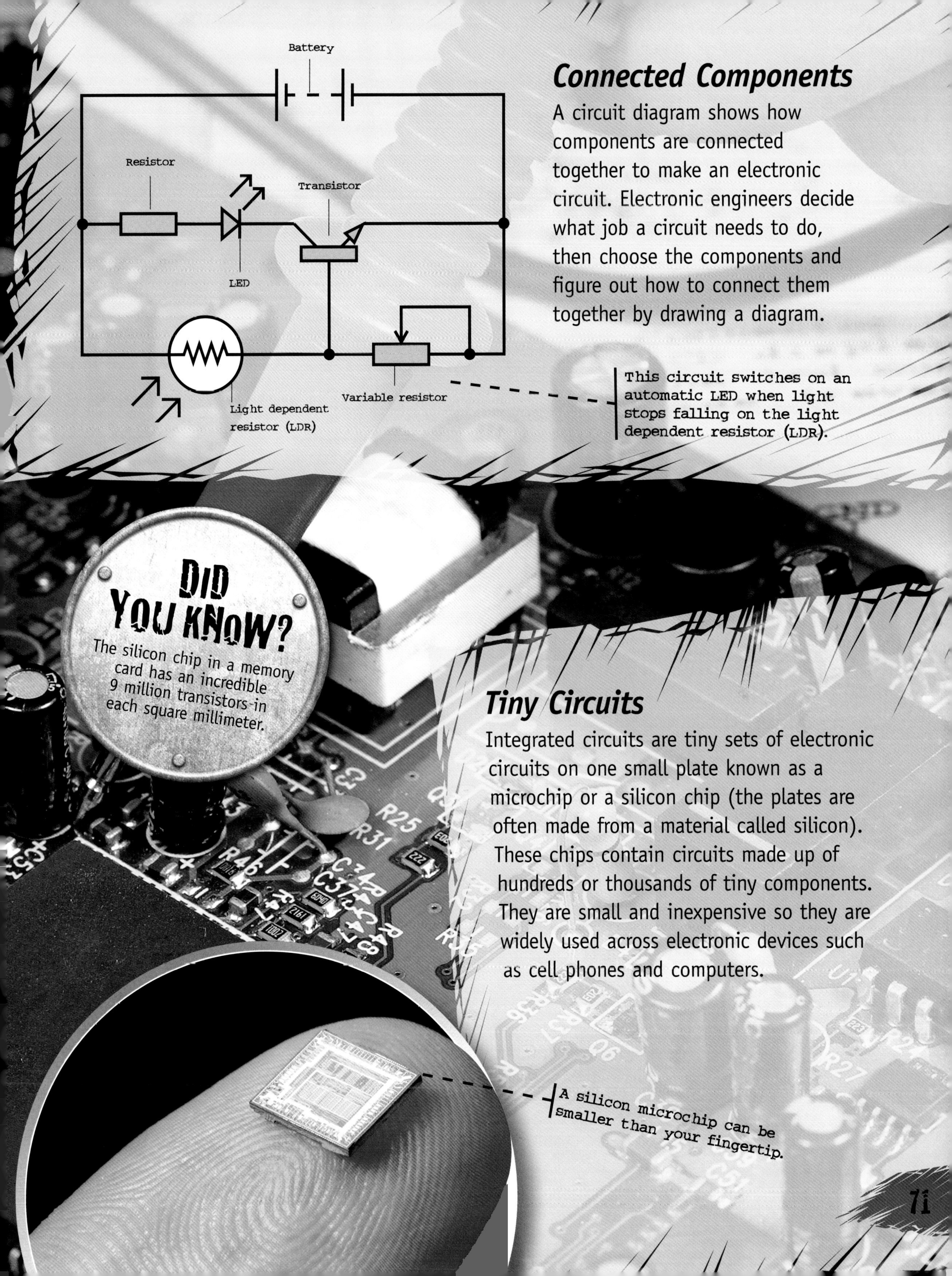
Battery
Resistor
Transistor
LED
Light dependent resistor (LDR)
Variable resistor
DID YOU KNOW?
The silicon chip in a memory card has an incredible 9 million transistors in each square millimeter.

Connected Components

A circuit diagram shows how components are connected together to make an electronic circuit. Electronic engineers decide what job a circuit needs to do, then choose the components and figure out how to connect them together by drawing a diagram.

This circuit switches on an automatic LED when light stops falling on the light dependent resistor (LDR).

Tiny Circuits

Integrated circuits are tiny sets of electronic circuits on one small plate known as a microchip or a silicon chip (the plates are often made from a material called silicon). These chips contain circuits made up of hundreds or thousands of tiny components. They are small and inexpensive so they are widely used across electronic devices such as cell phones and computers.

A silicon microchip can be smaller than your fingertip.

COMPUTERS

A computer is a complex electronic machine made up of two parts: hardware and software. Hardware refers to the electronic components and other parts of the actual computer. Software refers to the programs and data (information) stored in the machine.

DID YOU KNOW?
The most powerful supercomputers can do more than a quadrillion (a thousand million million) calculations every second.

Processor and Memory

A computer has a **processor,** which is its "brain" and memory, where programs and data are stored. The processor does calculations and also moves data from one part of the memory to another. Inside the processor, all data, from text to photographs, is stored in the form of binary numbers. These numbers are made up of the digits 1 and 0.

The main components of a computer, such as the processor and memory, are supported by the motherboard, the main circuit board found in computers.

Programming a Computer

A computer program (also called an application or app) is a list of instructions for a computer's processor to follow. Programs make a computer do different jobs, such as letting you play games, send e-mails, or browse the Internet.

Many computer programs are written in a language called Java.

Inputs and Outputs

Computers have inputs and outputs, where data enters and leaves the computer. A USB port on a personal computer is both an input and an output. You can input data into the port and the computer can also send data out of the port, to a printer for example.

USB flash drives are small, portable devices that can be used to store and transport data.

TABLETS

This is an Apple iPad tablet, with a screen of apps ready to be selected.

Home button. (When you press this, the iPad returns to the home screen.)

A tablet is a computer, just like a desktop computer or a laptop computer, but not as powerful. All the parts of a tablet computer are squeezed into a thin case, behind the tablet's touch screen.

Software apps

The icons (small pictures) on the screen are known as apps. Selecting an icon makes an app (which is a computer program) start working. Apps allow a tablet to do all sorts of different jobs, such as letting a user play games, send e-mails, navigate from place to place, or browse the Internet.

Tablet parts

The main parts of a tablet are the screen and the motherboard (this contains the tablet's processor and memory, the case, and the battery). There are also smaller parts, such as a camera, speakers, and a power socket, and various sensors such as tilt sensors and a GPS receiver.

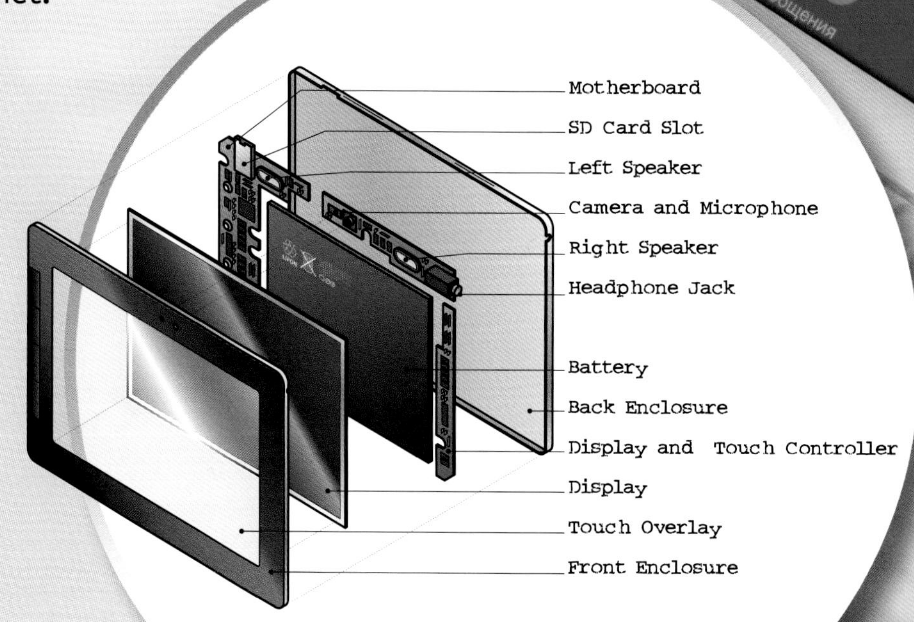

Touch screens

All tablets have a touch-sensitive screen (or touch screen). The top layer of the screen is a thin layer of glass that stores electricity. When you touch the screen, some of the electricity flows into your finger. Sensors at the edge of the screen detect this and figure out where you touched the screen. An app uses this information, for example, in a tic-tac-toe game, to determine which square on the board you have selected.

DID YOU KNOW?

The screen of an iPad display contains more than 3 million LCD pixels (see pages 94-95).

GAME CONSOLES

A game console is a computer that is used for playing games. The most important features of game consoles, such as the PlayStation 4 and the Xbox One, are super-fast graphics and special controllers.

Fast Graphics

Inside a game console are memory chips and a processor chip, as in all computers. A game console processor also contains a specialized electronic circuit called a graphics processing unit (GPU) that creates images on the screen. Where personal computers have a keyboard and mouse, or a track pad, a game console uses game controllers.

Drawing Pictures

The world of a computer game is made up of objects, such as trees, clouds, buildings, vehicles, and people or animals. The shape, size, color, texture, and position of these objects are stored in the console's memory. The processor uses a wire-frame made up of triangles, quadrilaterals, and other polygons to draw three-dimensional (3D) objects. There may be thousands of objects in the world, but the processor can use this method to draw any object dozens of times every second.

These wire-frame graphics show that objects are made up of many triangles and other polygons.

Game Control

Game consoles have special controllers that players use to control a game. Basic controllers have buttons and joysticks. Some consoles have gesture control, which means they can be controlled by hand movements, and even facial expressions. Game consoles have cameras and motion sensors that detect movement.

The Wii U is a home video game console. Its primary controller features an embedded touch screen.

DID YOU KNOW?

The Microsoft Xbox One's graphics processor can draw 1.6 billion colored triangles on screen in a second.

SIMULATORS

A simulator is a machine that creates pictures and sounds that look and sound like the real world. Some simulators create movement, too. Most simulators are flight simulators that pilots use to learn to fly before taking off in a real plane. There are also ship simulators and driving simulators.

Flight Simulator

The cabin of a flight simulator looks and feels exactly like the cockpit of a real plane. A powerful computer senses when the pilot moves the controls or presses switches, and figures out what would happen if it were a real plane. The computer also shows what it would look like outside the plane, and creates engine noise and other sounds so it is just like being in a real plane.

Making Movements

The flight simulator cabin is mounted on several **hydraulic rams**. The computer makes the arms move in and out, and this makes the cabin roll, tilt, and turn, as a real aircraft would, so it feels like flying.

A flight simulator cabin mounted on hydraulic rams.

Simulator Rides

Theme parks often have simulator rides that let you experience what it is like to fly in a jet fighter, go into space, hurtle along in a speedboat, or fly through the imaginary world of a movie. The riders sit in front of a huge screen watching a 3D image, on seats that tilt, vibrate, and move from side to side to match the action on the screen.

A Venturer motion ride simulator at a theme park.

This simulator shows an airport runway and is being used to train a pilot to land a plane.

DID YOU KNOW?

Flight simulators can help pilots train for emergencies, such as engine failures and landing without landing gear.

CELL PHONES

A cell phone is an electronic communications device. With any cell phone you can make calls and send text messages wherever you are. A smartphone is a tiny computer that has more features than a simple cell phone. It runs apps that allow it to do many different jobs.

DID YOU KNOW?

In a city center, there are dozens of mobile cells, each one covering just a few streets.

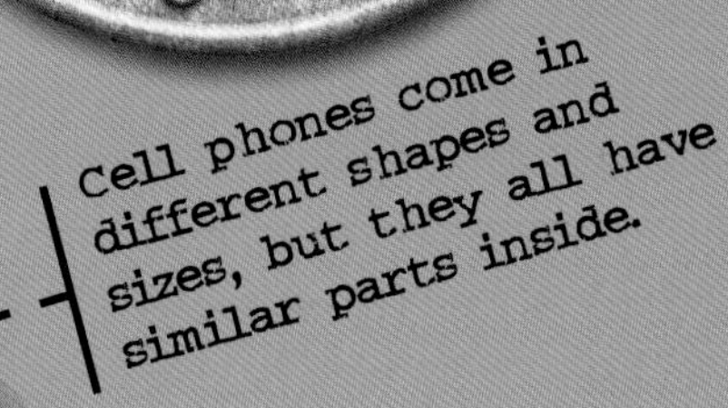

Cell phones come in different shapes and sizes, but they all have similar parts inside.

Parts of a Smartphone

There are lots of components crammed into the shell of a smartphone. It has a processor and a memory, which are attached to a circuit board. The aerial sends and receives the radio signals that the phone uses to communicate. The heaviest part of a smartphone is normally the battery.

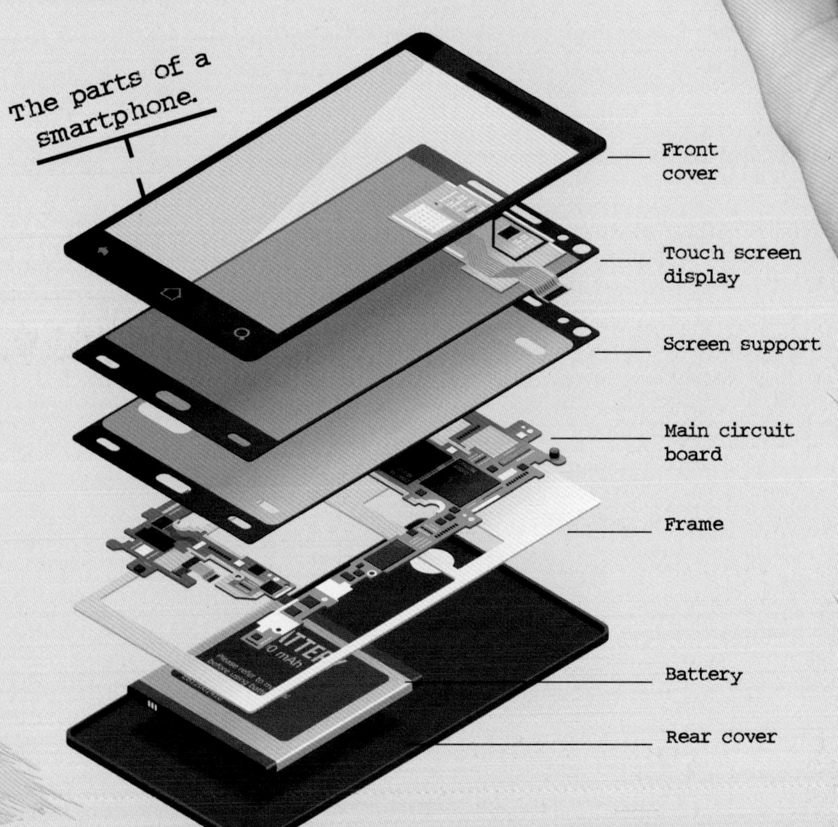

The parts of a smartphone.

Smartphone Apps

Apps allow smartphones to do many different jobs, such as playing games and finding directions, as well as making phone calls, taking pictures, and sending texts. E-mail apps, web browsing apps, social media apps, and many other apps need a connection to the Internet so that they can send and receive data.

The different apps on a cell phone are shown by various icons.

Mobile Cells

A cell phone connects to the cell phone network (and also to the Internet) by radio signals. A phone network has many aerials, and a cell phone automatically connects to the nearest one. The area around each aerial is called a cell. All the cells on a network are connected to one another and to other phone networks and the Internet. Phones need a small electronic card called a SIM card that allows them to connect to a cell network.

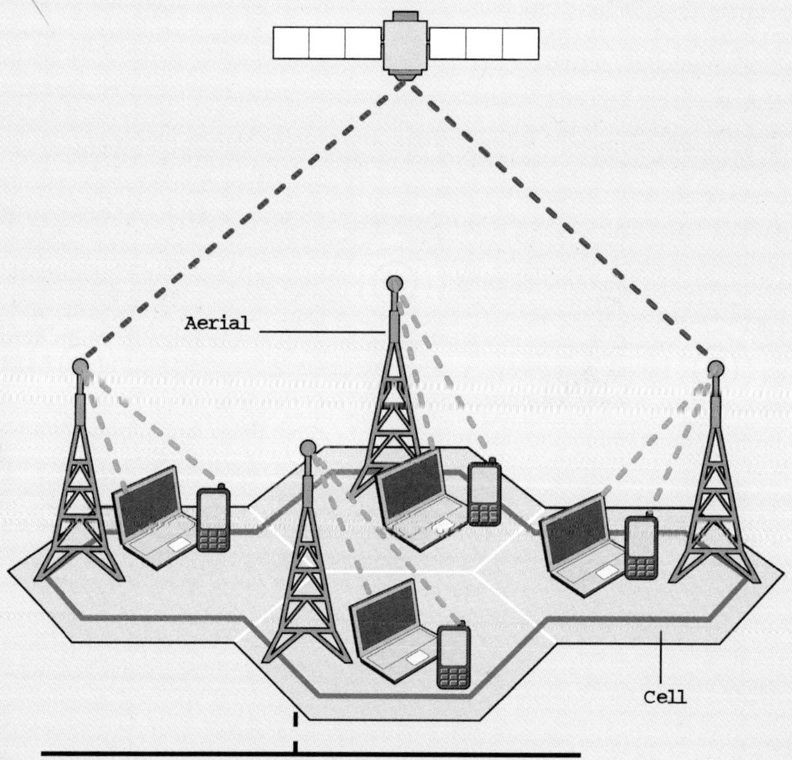

Phones connect to aerials in areas called cells. The cells are connected to one another by radio links..

CAMERAS

Before digital cameras, film was used to capture images. The film had to be processed using special chemicals to print photographs on paper. Today, a digital camera can create images and make videos at the touch of a button. These can be viewed on a screen, printed, or shared online.

Parts of a Digital Camera

A digital camera has a lens, an image sensor, a monitor for composing and viewing photographs and videos, and a memory card for storing photographs and video. The lens bends rays of light from the subject so that the light makes a small picture of the scene.

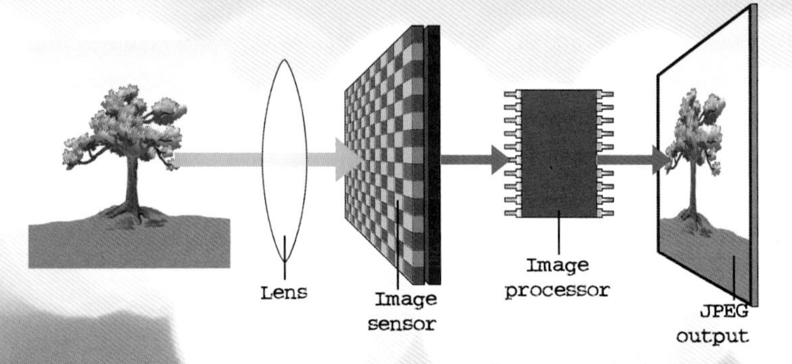

A lens bends light to make an image of the subject on the image sensor.

Recording the Image

An image sensor picks up the information that will create the photograph. The face of the image sensor is divided into thousands of tiny squares called pixels. The light from the object being photographed changes the amount of electricity in each pixel, and a sensor measures how much electricity is left. The camera's image processor takes these measurements from the sensor and figures out the the color and brightness of each pixel in the picture, and stores the picture on the camera's memory card.

An image sensor lives inside the camera.

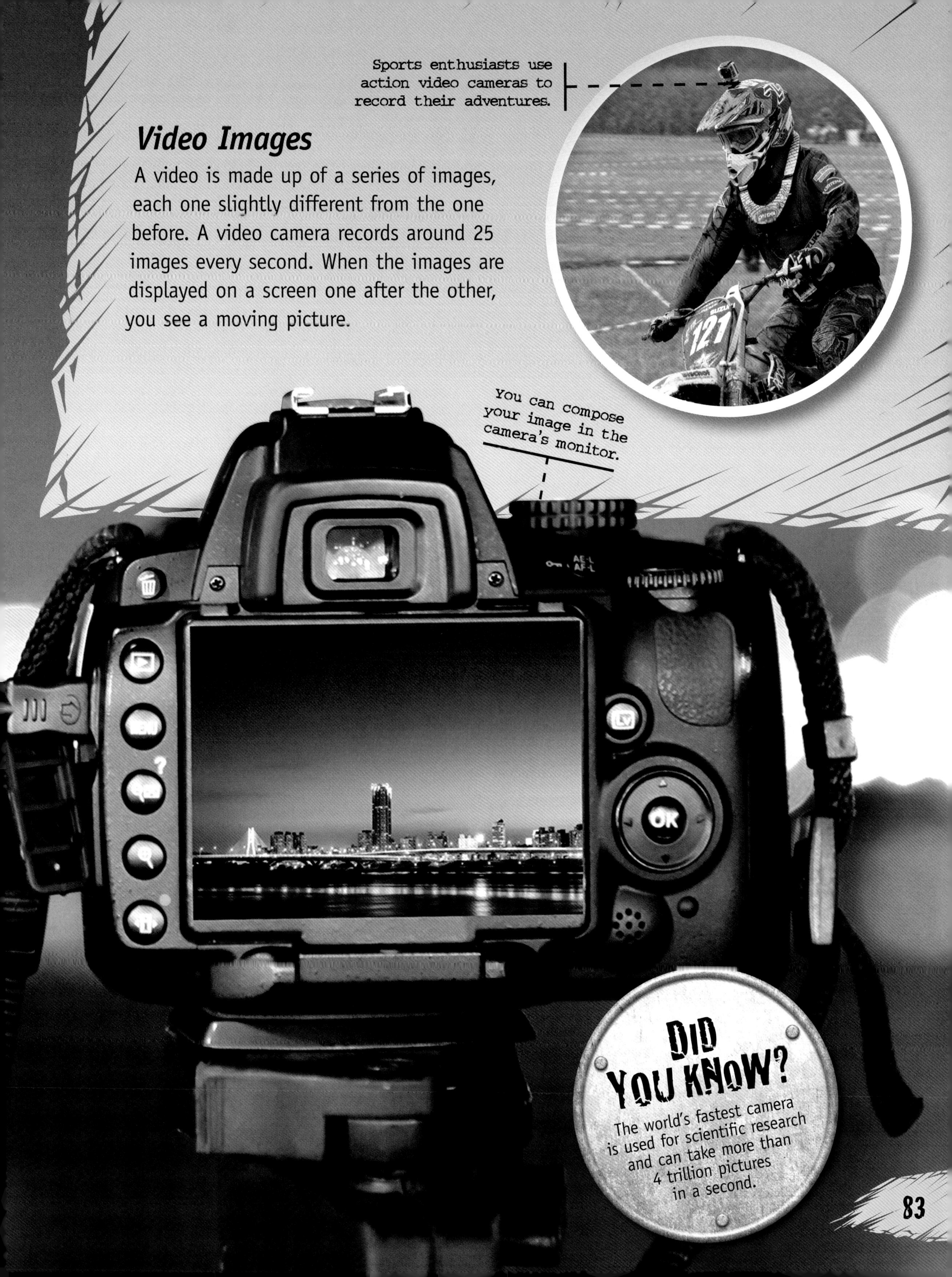

Sports enthusiasts use action video cameras to record their adventures.
You can compose your image in the camera's monitor.
DID YOU KNOW?
The world's fastest camera is used for scientific research and can take more than 4 trillion pictures in a second.

Video Images

A video is made up of a series of images, each one slightly different from the one before. A video camera records around 25 images every second. When the images are displayed on a screen one after the other, you see a moving picture.

TELECOMMUNICATION

The word telecommunication means communicating over a distance using electricity. Telephones, radio, and television are all types of telecommunication. A telecommunications network allows devices around the world to connect to one another.

Traveling Signals

Sound from telephones, and data from computers, is turned into electrical signals before it goes into a telecommunications network. These signals travel along wires. In some parts of a network, the signals are turned into radio waves which are beamed between aerials. In other parts, they are turned into light signals via optical links.

Optical Links

Many links in telecommunication networks use fiber optic cables. A cable contains dozens of tiny glass fibers. Electrical signals are turned into flashes of light, which are fired down the fibers. At the other end of the cable, the flashes are turned back to electrical signals.

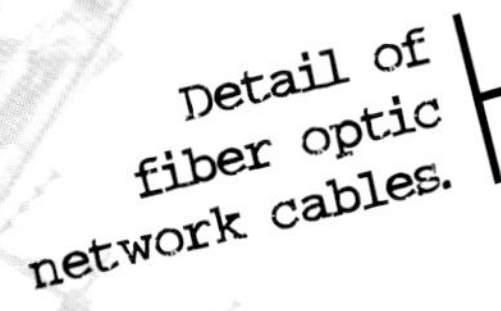

Detail of fiber optic network cables.

DID YOU KNOW?
Signals travel along fiber optic cables at the speed of light—that's 186,000 miles a second.
Telecommunications towers and masts are very tall so that radio signals can travel between them without trees, buildings, or hills blocking the signals.

Making a Connection

All phones are connected to switching centers that are part of the phone network, and these switching centers are all connected to one another. As you dial a number, the switching centers set up a connection from your phone to the phone you are calling.

THE INTERNET

The Internet is a vast computer network that allows us to browse websites and send e-mails (see pages 88–89), download games, watch videos, and do many other tasks. The Internet is made up of many smaller computer networks, each made up of computers, tablets, and smartphones that can exchange information with one another.

The Internet connects people all around the world.

Internet Connections

Parts of the Internet in different areas of the world are connected by high-speed, high capacity fiber optic cables, which make up the Internet's "backbone." Many of these cables go under seas and oceans. They allow huge amounts of data to travel from one side of the world to the other in fractions of a second.

A Small Network

Most computers and devices are connected to the Internet in small networks. These networks connect to the Internet's "backbone" through an Internet service provider (ISP). Networks often contain computers called servers. Other computers on the Internet, anywhere in the world, can ask servers for data, and send information to servers.

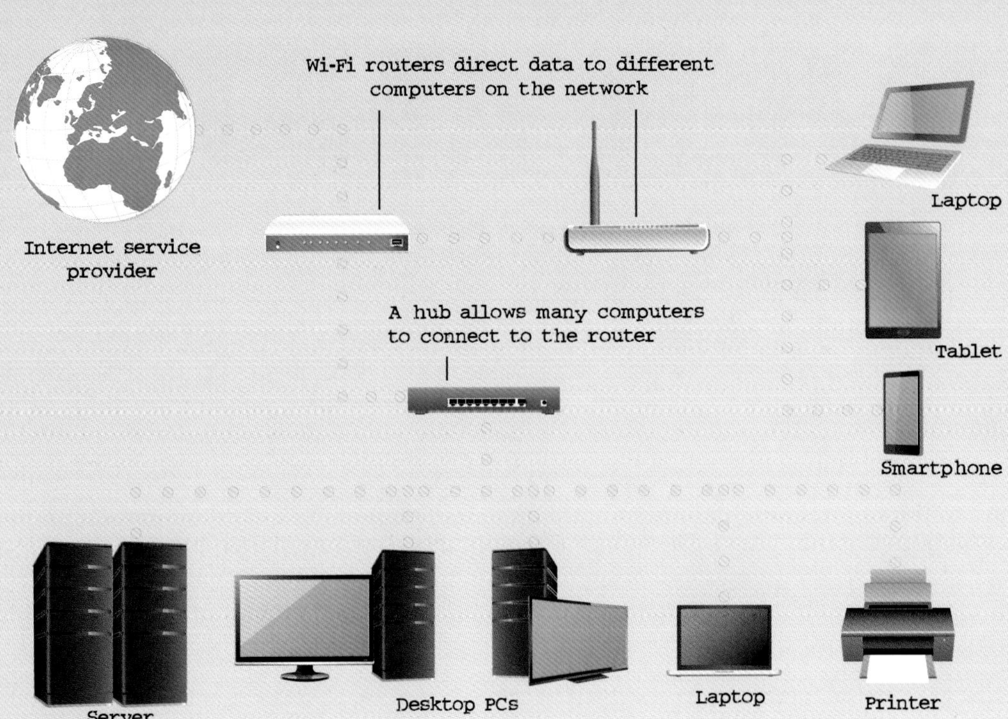

An example of a network of computers connected to a server.

Packets of data can reach your device from the other side of the world in less than a second.

Packets of Data

When data, such as a music file, moves around the Internet, it is divided into small chunks called packets. Each computer on the Internet has its own number, called an Internet Protocol (IP) address. Packets are labeled with the IP address so they know where to go. Computers called routers read the address to make sure that packets are sent to the right place.

DID YOU KNOW?

The Internet was first conceived in the 1960s-1970s but wasn't in wide use until the 1990s. Now, two billion people worldwide use the Internet.

WEB AND E-MAIL

The World Wide Web (WWW) is a vast collection of information stored on computers around the world that you can explore through the Internet. E-mail is sending and receiving messages between computers, tablets, and smartphones on the Internet.

DID YOU KNOW?
More than 200 million e-mails are sent through the Internet every minute.

Web Pages and Sites

A web page is a piece of information stored on a server on the Internet. A website is a collection of web pages. A web page has links on it that you click to move to other pages on the same website, or to other websites. Every web page has its own address (called a URL) that you use to find it.

URL bar

Every web page has i
own address located
the URL bar.

The World Wide Web gives us access to all sorts of information and entertainment from all over the world.

Website Code

A web page is stored in a computer code called hypertext mark-up language (HTML). This code contains all the information needed to display the website. When you put the address of a website into a web browser, the browser gets the HTML code from the server where the web page is stored, and then draws the web page on screen.

Sending Messages

E-mail is short for electronic mail. To send and receive e-mails, you need an application called an e-mail client. This sends your messages into the Internet, and collects messages that have been sent to you. Popular email clients include Apple Mail and Microsoft Outlook. Online applications such as Gmail, Yahoo, or Hotmail also act as e-mail clients. An e-mail address is made up of a user name, then an @ sign (which means "at"), then the name of the server where the message will be stored.

Your email client manages your Inbox (incoming mail) and Outbox (outgoing mail).

WI-FI AND BLUETOOTH

Wi-Fi and Bluetooth are two ways that computers, tablets, mobile phones, and other devices that are near to one another can exchange data. Wi-Fi and Bluetooth use invisible radio waves, instead of wires or cables, to connect devices. This is called wireless technology.

Wi-Fi Networks

Wi-Fi is used to connect computers, smartphones, and tablets to the Internet and to one another. Radio waves connect the machines to a box called a Wi-Fi router. This is called a Wi-Fi network.

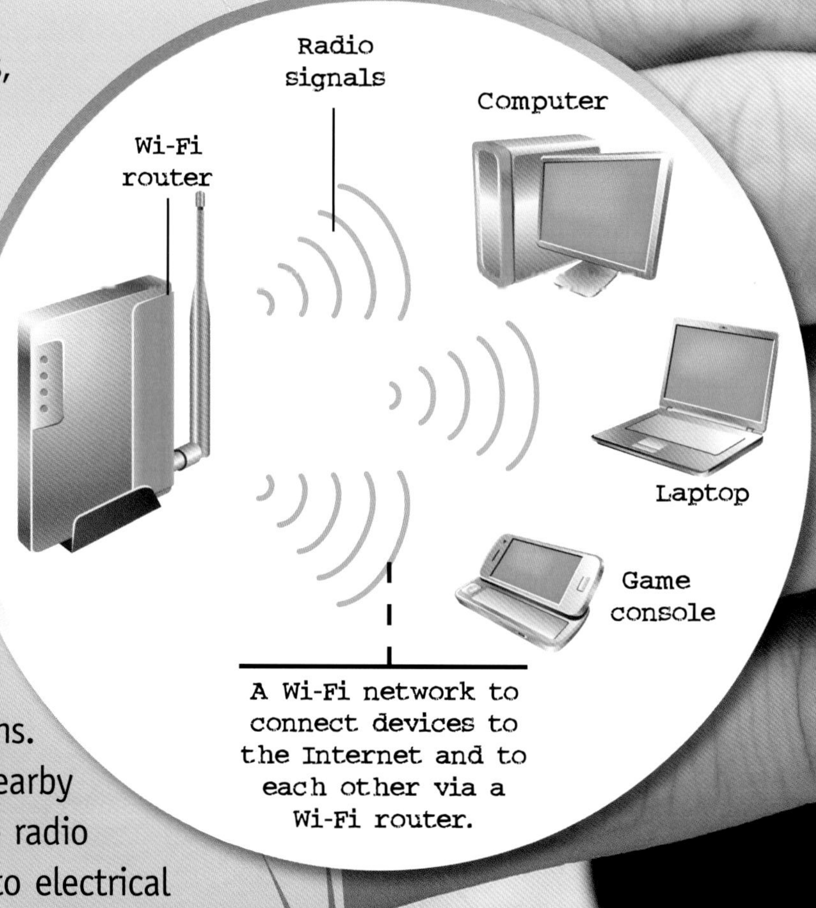

A Wi-Fi network to connect devices to the Internet and to each other via a Wi-Fi router.

Radio Connections

A Wi-Fi router has an aerial that turns electrical signals into radio signals which are then transmitted in all directions. Computers and other devices nearby use their aerials to pick up the radio signals and turn them back into electrical signals. The computers and other devices can send data to the router in the same way.

Bluetooth Connections

Bluetooth is used to connect mobile phones to headsets, keyboards and mice to computers, game controllers to game consoles, and radio-controlled toys to tablets or smartphones. Before two devices can exchange data, one of the devices sends a message to the other to say that it wants to connect. If the other device sends a reply, the connection is made. Making this connection is known as pairing.

This man is using Bluetooth to link his headset and smartphone.

These two smartphones are connected by a Bluetooth link.

DID YOU KNOW?

Wi-Fi networks have passwords so that you can stop people from joining your network without asking.

RADIO AND BROADCASTING

Radio waves are invisible waves that travel through the air and through space. We use radio waves for many sorts of communications, such as mobile phones, walkie-talkies, radio broadcasting, and Wi-Fi networks.

Shaping Radio Waves

We transmit (send) sounds and other information using shaped radio signals. For example, at a radio station, a plain electrical signal is fed to the radio transmitter, where it is shaped by a signal from a microphone. The transmitter then turns the shaped signal into a radio signal that spreads out from the transmitter. A radio receiver detects the radio signal and turns it back into sound.

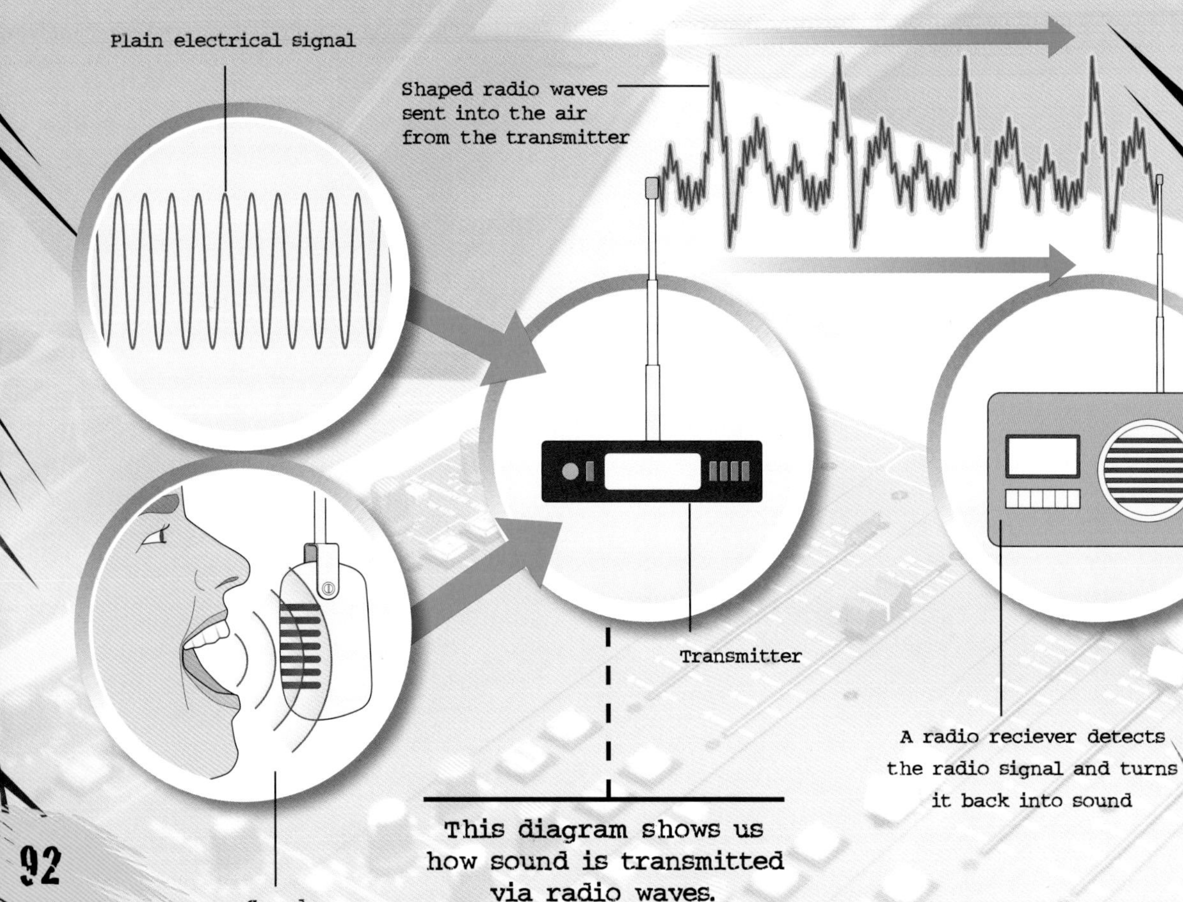

This diagram shows us how sound is transmitted via radio waves.

An aerial converts electric power into radio waves, and vice versa.
DID YOU KNOW?
Radio signals travel at the speed of light—that's 186,000 miles per second.
There are more than 13,500 radio stations broadcasting across the USA.

One-to-one Radio

When you speak into a walkie-talkie, the walkie-talkie sends out radio waves that are shaped to represent the sound of your voice. The receiver of the other walkie-talkie handset detects the signal, and the walkie-talkie uses the shape of the signal to rebuild the sound of your voice.

Broadcasting

Broadcasting means sending a radio signal from a transmitter to many different receivers at the same time. In radio broadcasting, the signals carry sounds from a radio studio to radio receivers. In television broadcasting, the signals carry television pictures to television receivers. The signals can be transmitted from aerials on Earth's surface, or from satellites in space.

SCREENS AND DISPLAYS

Screens and other displays show us images produced by computers, tablets, phones, cameras, calculators, watches, and many other devices. Screens and displays work by producing and controlling light.

Adding Colors

The color screens in most devices and televisions produce colors using tiny red, blue, and green dots of light. Different colors are made by changing the brightness of the different lights. For example, turning up red and blue, and turning down green, creates purple light. A screen contains hundreds of thousands of colored dots. The electronic circuits that operate the screen adjust the brightness of each dot to make the picture.

DID YOU KNOW?

The largest jumbo screen in the world, in Suzhou, China, is 1,649 feet long and contains two million LEDs.

Times Square in New York City is filled with animated LED signs.

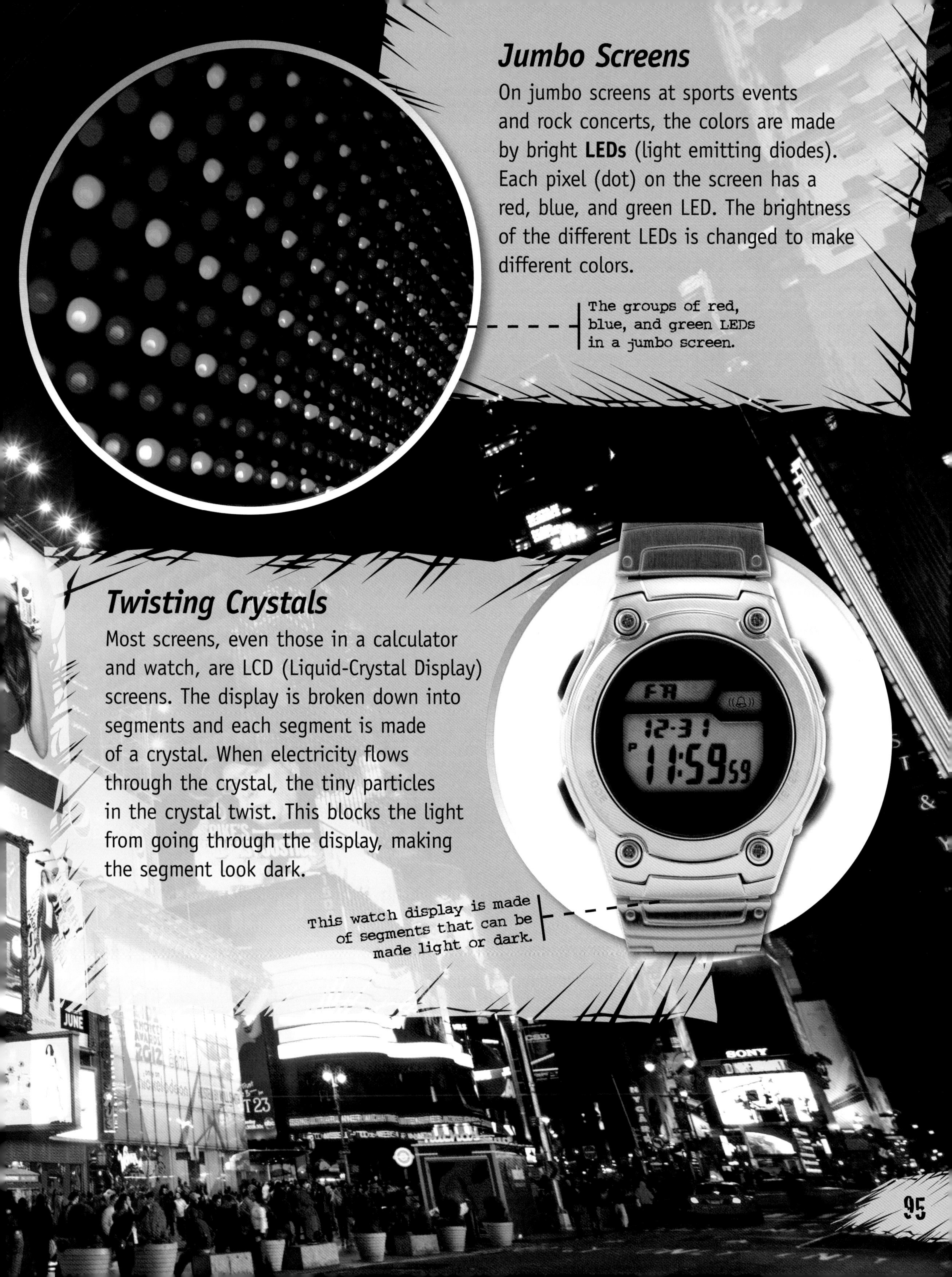

Jumbo Screens

On jumbo screens at sports events and rock concerts, the colors are made by bright **LEDs** (light emitting diodes). Each pixel (dot) on the screen has a red, blue, and green LED. The brightness of the different LEDs is changed to make different colors.

The groups of red, blue, and green LEDs in a jumbo screen.

Twisting Crystals

Most screens, even those in a calculator and watch, are LCD (Liquid-Crystal Display) screens. The display is broken down into segments and each segment is made of a crystal. When electricity flows through the crystal, the tiny particles in the crystal twist. This blocks the light from going through the display, making the segment look dark.

This watch display is made of segments that can be made light or dark.

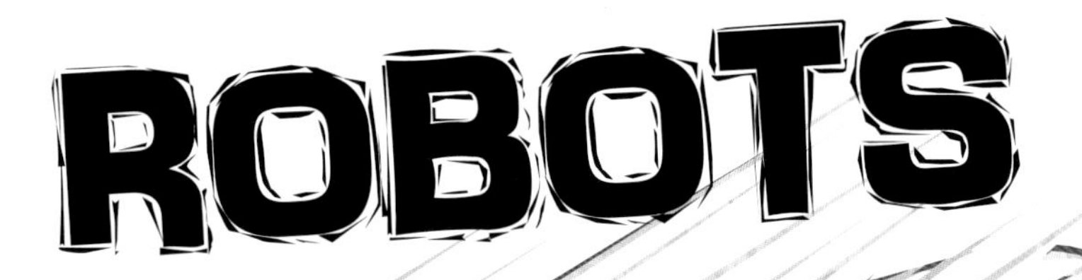

ROBOTS

A robot is a machine designed to do jobs for us. Most robots are industrial robots that work in factories, but scientists are also building experimental robots that move like humans and animals. Some submersibles (see page 38), drones (see page 50), and space probes (see page 68) are also robots.

DID YOU KNOW?

Armies use robots to check out enemy territory and clear landmines.

Industrial robots weld together parts of cars.

Robot Workers

Factory robots are used to move parts and materials from place to place, and are used to operate tools. Most factory robots are robotic arms. The arm is like a human arm, with a shoulder, elbow, and wrist that can bend and turn. Tools such as grippers can hold objects, welding tools, and screwdrivers. Paint sprayers can also be attached to the arm. A computer controls the robot, making it repeat the same movements again and again.

Walking Robots

Building a robot that can walk or run without falling over is very difficult. Complex computer programs are needed to control walking robots, and the robots have tilt and touch sensors that feed back information to the programs.

A Honda "ASIMO" humanoid robot. ASIMO stands for "Advanced Step in Innovative Mobility."

Human-like robotic hands need motors to bend each finger and touch sensors on the fingertips.

Movement and Sensors

The parts of robots, such as arms and grippers, are moved by electric motors or hydraulic rams. Robots have sensors that send information to the computers that control them. For example, touch sensors in a gripper tell the computer when the gripper has picked up an object.

BODY SCANNERS

Doctors often need to take pictures of the insides of patients' bodies to check for problems such as broken bones. They do this with X-ray machines and with medical scanners such as MRI scanners, CT scanners, and ultrasound scanners.

A patient being scanned inside an MRI scanner.

An X-ray is good for seeing bones in the body.

X-rays

X-rays are invisible rays, like radio waves and microwaves. They pass through soft materials, such as muscles, but not hard materials, such as bones. An X-ray machine fires X-rays through a patient's body. An X-ray film or sensor on the other side detects the rays that get through.

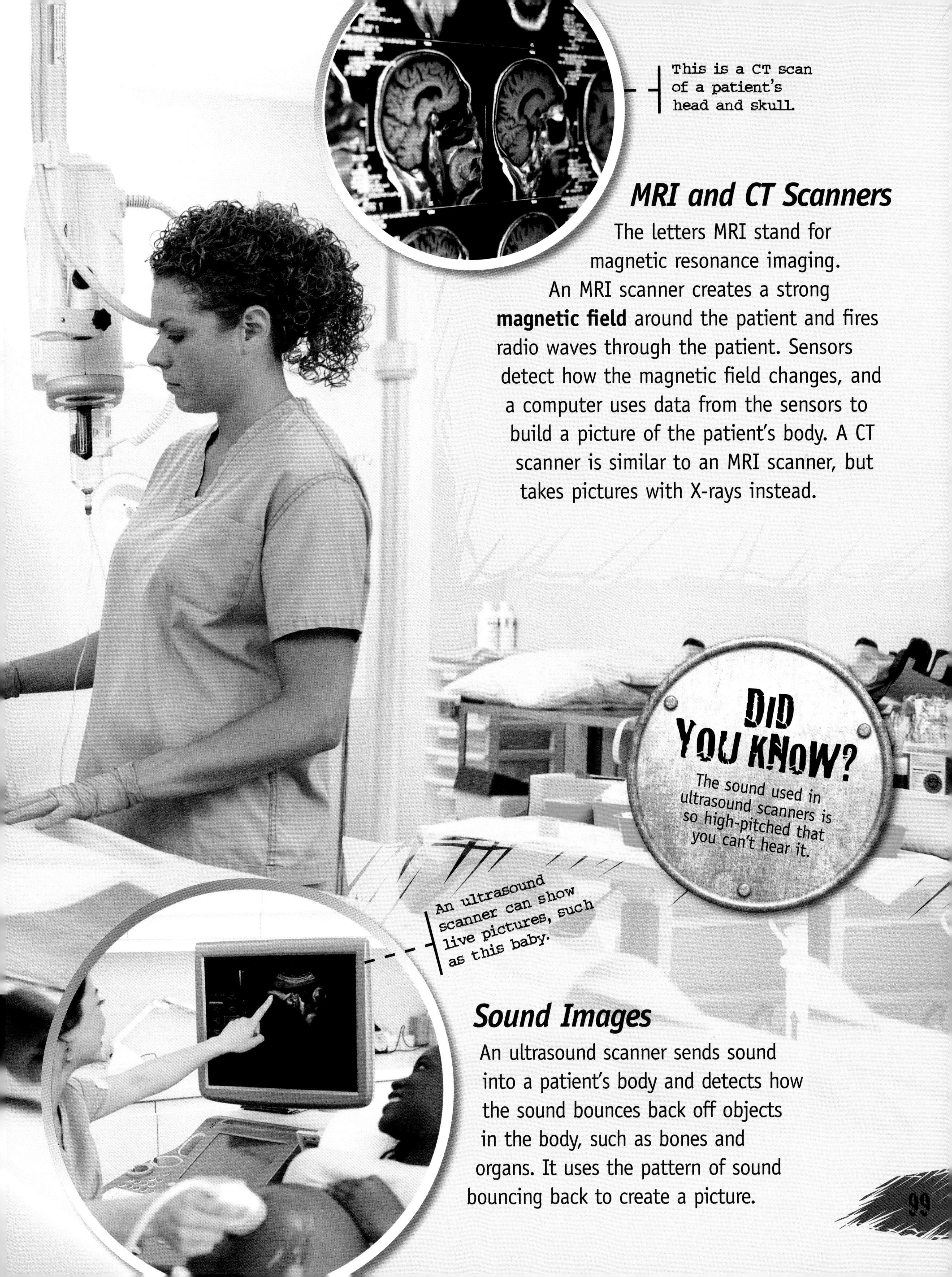

This is a CT scan of a patient's head and skull.

MRI and CT Scanners

The letters MRI stand for magnetic resonance imaging. An MRI scanner creates a strong **magnetic field** around the patient and fires radio waves through the patient. Sensors detect how the magnetic field changes, and a computer uses data from the sensors to build a picture of the patient's body. A CT scanner is similar to an MRI scanner, but takes pictures with X-rays instead.

DID YOU KNOW?

The sound used in ultrasound scanners is so high-pitched that you can't hear it.

An ultrasound scanner can show live pictures, such as this baby.

Sound Images

An ultrasound scanner sends sound into a patient's body and detects how the sound bounces back off objects in the body, such as bones and organs. It uses the pattern of sound bouncing back to create a picture.

TELESCOPES

A telescope is an instrument that makes objects far away look larger, so you can see more detail. Astronomers use telescopes to study objects in space, and people use telescopes to watch wildlife and sports.

Yerkes Observatory is operated by the University of Chicago, Wisconsin. It has the world's largest refracting telescope.

Optical Telescopes

Most telescopes are called optical telescopes because they collect light coming from distant objects. The main types of optical telescopes are refracting telescopes and reflecting telescopes. A refracting telescope collects light with a large lens, and a reflecting telescope collects light with a large, curved mirror.

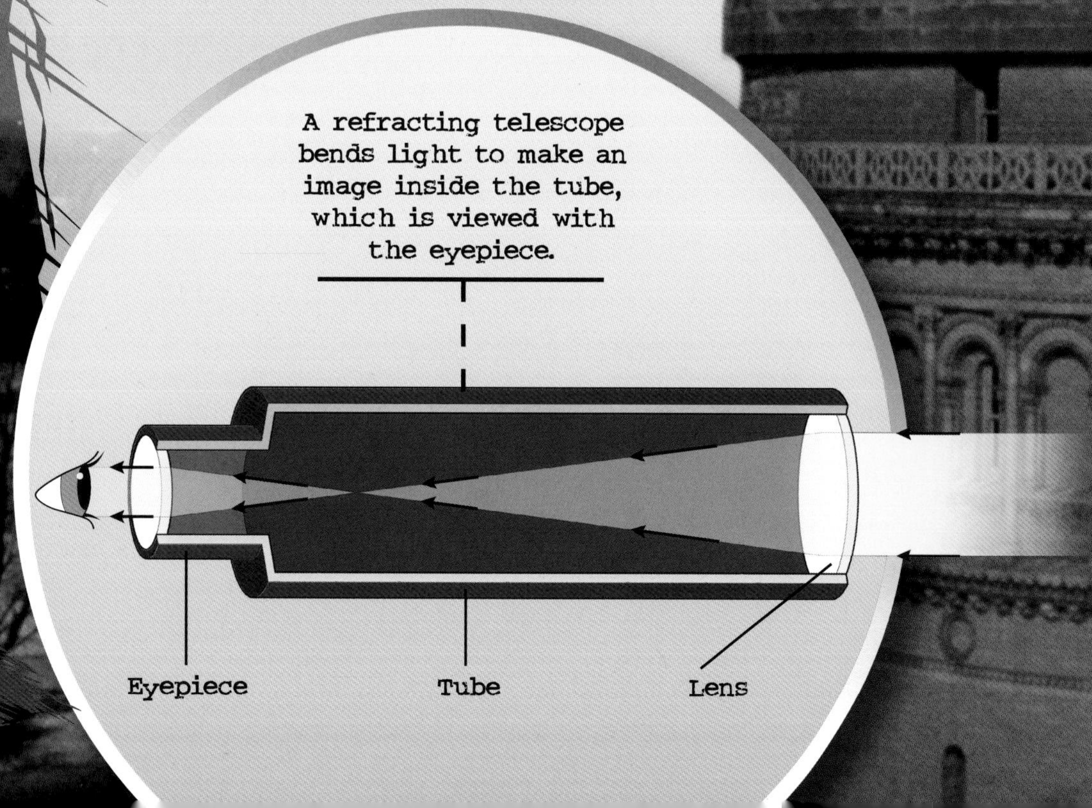

A refracting telescope bends light to make an image inside the tube, which is viewed with the eyepiece.

Eyepiece

Tube

Lens

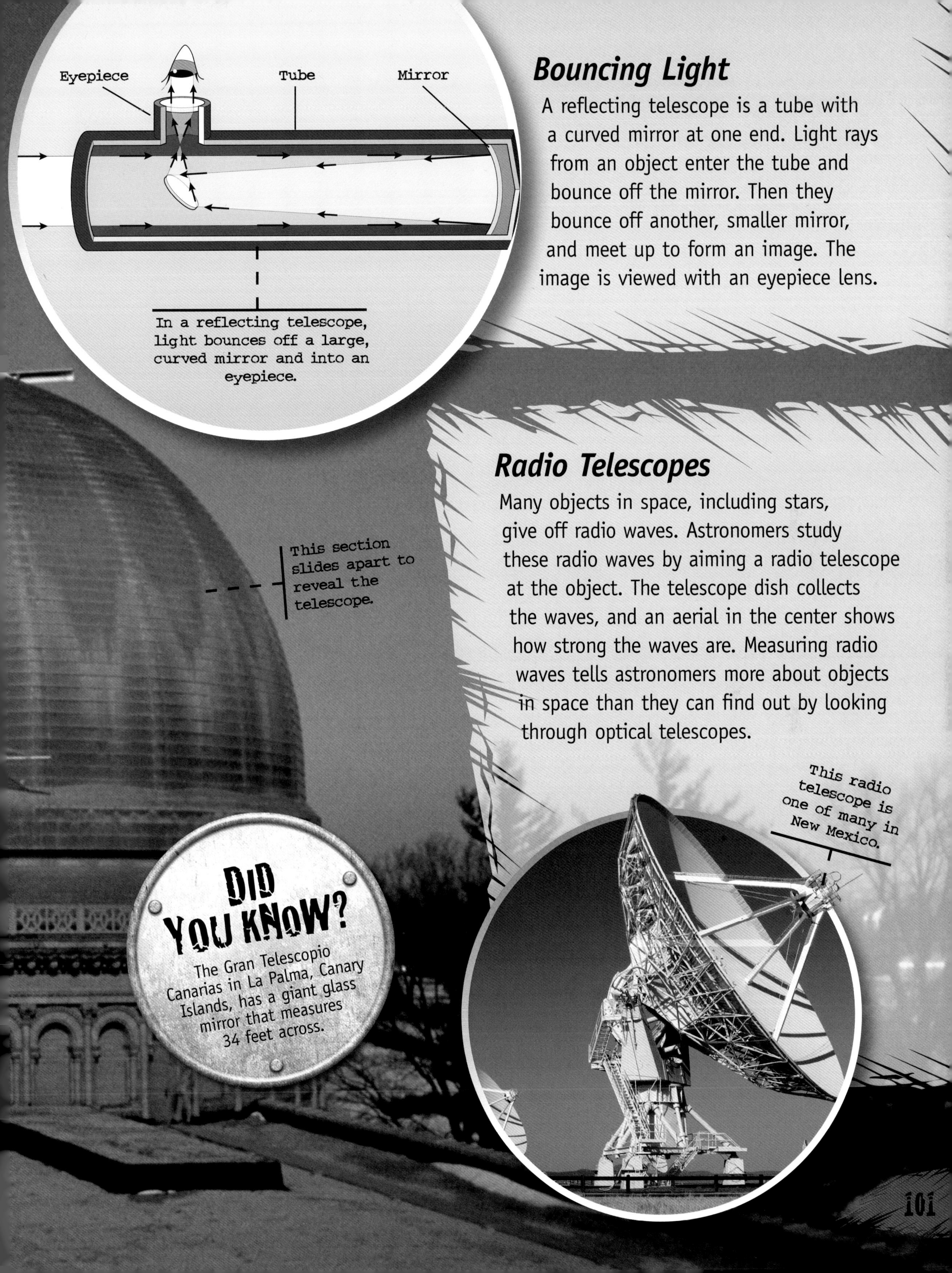

Eyepiece
Tube
Mirror
This section slides apart to reveal the telescope.
This radio telescope is one of many in New Mexico.

In a reflecting telescope, light bounces off a large, curved mirror and into an eyepiece.

Bouncing Light

A reflecting telescope is a tube with a curved mirror at one end. Light rays from an object enter the tube and bounce off the mirror. Then they bounce off another, smaller mirror, and meet up to form an image. The image is viewed with an eyepiece lens.

Radio Telescopes

Many objects in space, including stars, give off radio waves. Astronomers study these radio waves by aiming a radio telescope at the object. The telescope dish collects the waves, and an aerial in the center shows how strong the waves are. Measuring radio waves tells astronomers more about objects in space than they can find out by looking through optical telescopes.

DID YOU KNOW?

The Gran Telescopio Canarias in La Palma, Canary Islands, has a giant glass mirror that measures 34 feet across.

MICROSCOPES

A microscope is an instrument that makes tiny objects look larger. It magnifies the objects so we can study them. There are two main types of microscopes: optical microscopes and electron microscopes.

Electron microscopes are large, expensive pieces of equipment.

Optical Microscopes

Optical microscopes make images using light. The object to be studied is put on the microscope's stage. Light from a light source goes up through the object and into a lens, which makes an image of the object. The image is viewed through an eyepiece or by a camera.

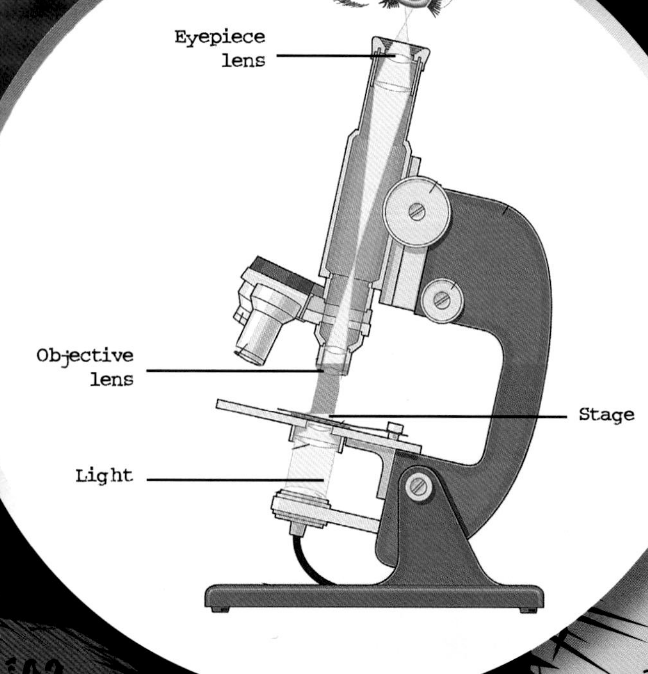

Bending Light

An optical microscope works by bending light with lenses. Light rays from the object go into an objective lens at the bottom of a tube. The lens bends the light rays so that they come together near the top of the tube, making an image. The image is viewed through an eyepiece lens.

Electron Microscope

An electron microscope uses tiny particles called electrons to make an image. Electron microscopes are able to magnify objects much more than an optical microscope, allowing us to see much smaller objects in detail. Scanning Electron Microscope (SEM) sensors detect how the electrons bounce off the object, and the microscope uses these signals to build up a magnified 3D image of the object. A Transmission Electron Microscope (TEM) fires electrons through a thin slice of an object and makes 2D images, like the images from an optical microscope.

The components of a scanning electron microscope (SEM)

LASERS

A laser is a device that produces a stream of bright light called a laser beam. Lasers have many different uses, including light shows, cutting machines, laser surgery, barcode scanning, and measuring distances.

Laser Beams

Light rays always travel in straight lines unless they get reflected by mirrors or bent by lenses. Laser beams are made up of many light rays all traveling together in the same direction, which makes them narrow and very bright.

Laser beams fired into the sky create spectacular light displays.

Laser Cutting

Lasers that produce very powerful laser beams are often used in factories for cutting metal very accurately. The beam is focused onto the metal with a lens. The laser heats the metal very quickly, making it melt almost instantly. In medicine, surgeons often use low-power lasers instead of sharp knifes for cutting during operations.

Cutting shapes from metal sheets with a laser cutting tool.

Laser light is produced by atoms in a laser tube.

Making a Laser Beam

There are several different types of laser, but they all work in a similar way. A basic laser is a tube. A light source gives energy to the atoms in the tube, so the atoms move around a lot and by doing this, they produce light. Mirrors at either end of the tube trap the light so it bounces back and forth through the tube. This makes the atoms move even more, creating more energy and light, until there is enough energy to break out through the mirror as a strong laser beam.

DID YOU KNOW?

Scientists have accurately measured the distance to the Moon by firing a laser beam at a reflector left on the Moon by Apollo astronauts.

THE LARGE HADRON COLLIDER (LHC)

The Large Hadron Collider (LHC) is a giant scientific instrument called a particle accelerator. Scientists are using the LHC to find out more about the tiny particles that make up everything in the universe.

The LHC's main underground tunnel is almost 17 miles long.

Underground Tunnel

The LHC is a system of underground tunnels, including a large main tunnel. Billions of tiny particles smaller than atoms travel around in one direction and other particles travel in the opposite direction so that they collide. Detectors in the tunnels look for new particles made when they collide.

Powerful Magnets

The particles travel through 1,740 super-powerful magnets along the tubes. The magnets make the particles travel faster and faster. The particles almost reach the speed of light (186,000 miles per second). They hurtle around the main tunnel 11,000 times a second.

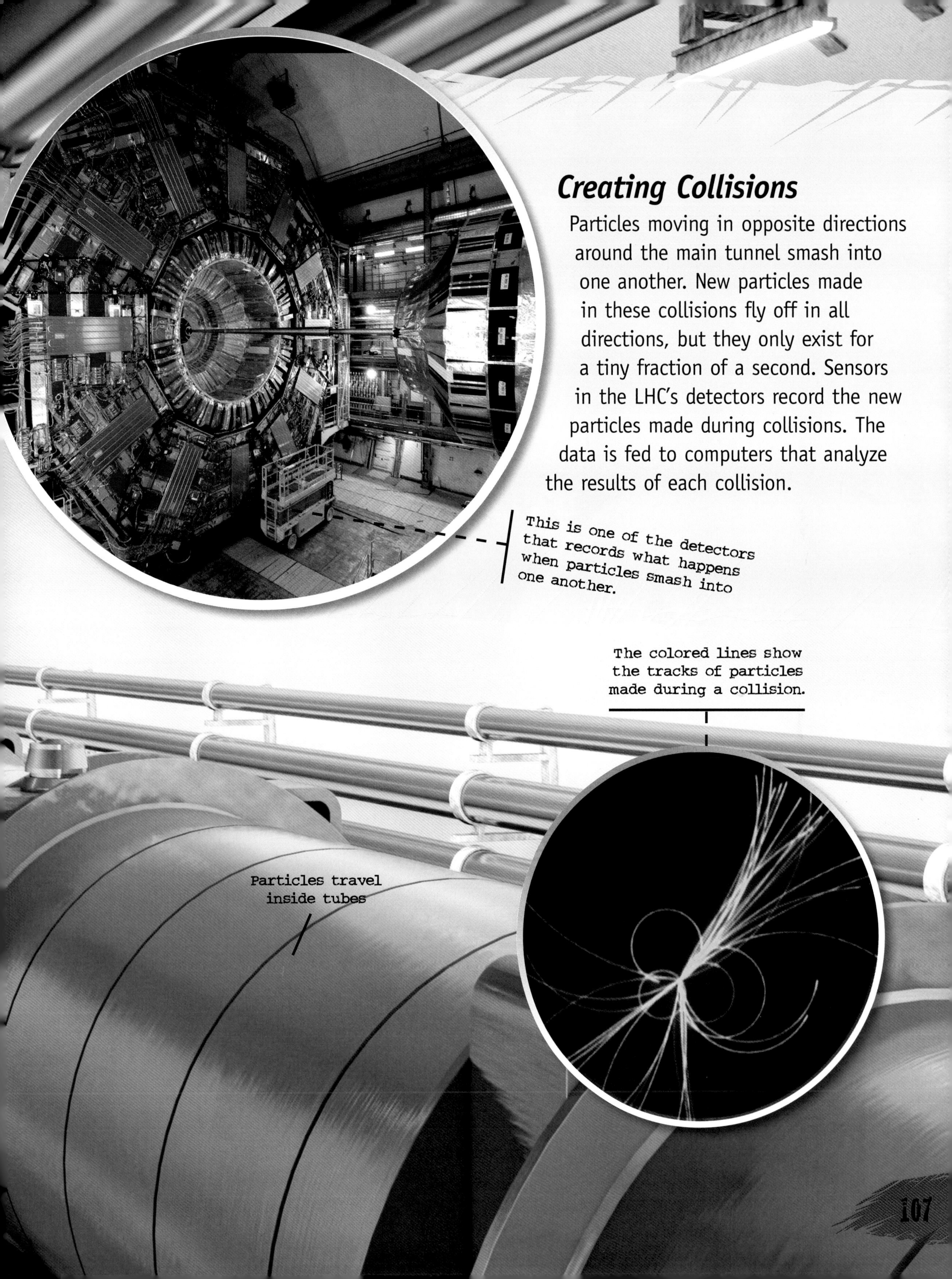

Creating Collisions

Particles moving in opposite directions around the main tunnel smash into one another. New particles made in these collisions fly off in all directions, but they only exist for a tiny fraction of a second. Sensors in the LHC's detectors record the new particles made during collisions. The data is fed to computers that analyze the results of each collision.

This is one of the detectors that records what happens when particles smash into one another.

The colored lines show the tracks of particles made during a collision.

Particles travel inside tubes

SKYSCRAPERS

A skyscraper is a very tall building with dozens of stories. Skyscrapers are built to create office and living space in crowded cities, where there's very limited space on the ground. Some clever engineering is needed to prevent these amazing buildings from toppling over.

Skyscrapers like these in Manhattan are used as offices, shops, and homes.

Frames, Floors, and Walls

The part of a skyscraper that is above the ground is called the superstructure. The superstructure is made up of immensely strong steel or a reinforced concrete frame (see pages 110-111). The frame supports the floors. Thin walls, called curtain walls, hang from the building's frame like curtains.

The Petronas Towers in Kuala Lumpar are twin towers and each tower is 1,483 feet high.
DID YOU KNOW?
In 2010, the Burj Khalifa opened in Dubai. At 2,716 feet tall, it is the world's tallest skyscraper.
Concrete slab
Piles sit on a specially constructed concrete slab

Foundations

A skyscraper's superstructure is supported by foundations. The foundations stop the building from sinking into the ground, and also hold the building upright, working like the roots of a tree. The superstructure sits on concrete and steel legs, called piles.

BUILDING SKYSCRAPERS

The construction of a new skyscraper starts with an architect, who designs the building, working out its shape, structure, and what materials will be used. Foundations are placed first, then the frame is constructed on top, before floors and walls are added.

New construction of a high-rise concrete-framed building in Burnaby, Canada.

Steel frames are erected using cranes.

Steel Frame

Steel is also used to make foundations and frames for skyscrapers. Steel pile foundations are driven into the ground by a machine called a piledriver. A steel frame comes in hundreds or thousands of columns and beams, which are bolted or welded together to make the frame.

Concrete Structures

Skyscrapers need strong foundations and frames to support their floors and walls. Some of these foundations and frames are made from reinforced concrete—concrete with steel bars running through it. Pile foundations are made by digging shafts and filling them with reinforced concrete. A concrete frame is made by pouring wet concrete into molds that make up each part of the frame.

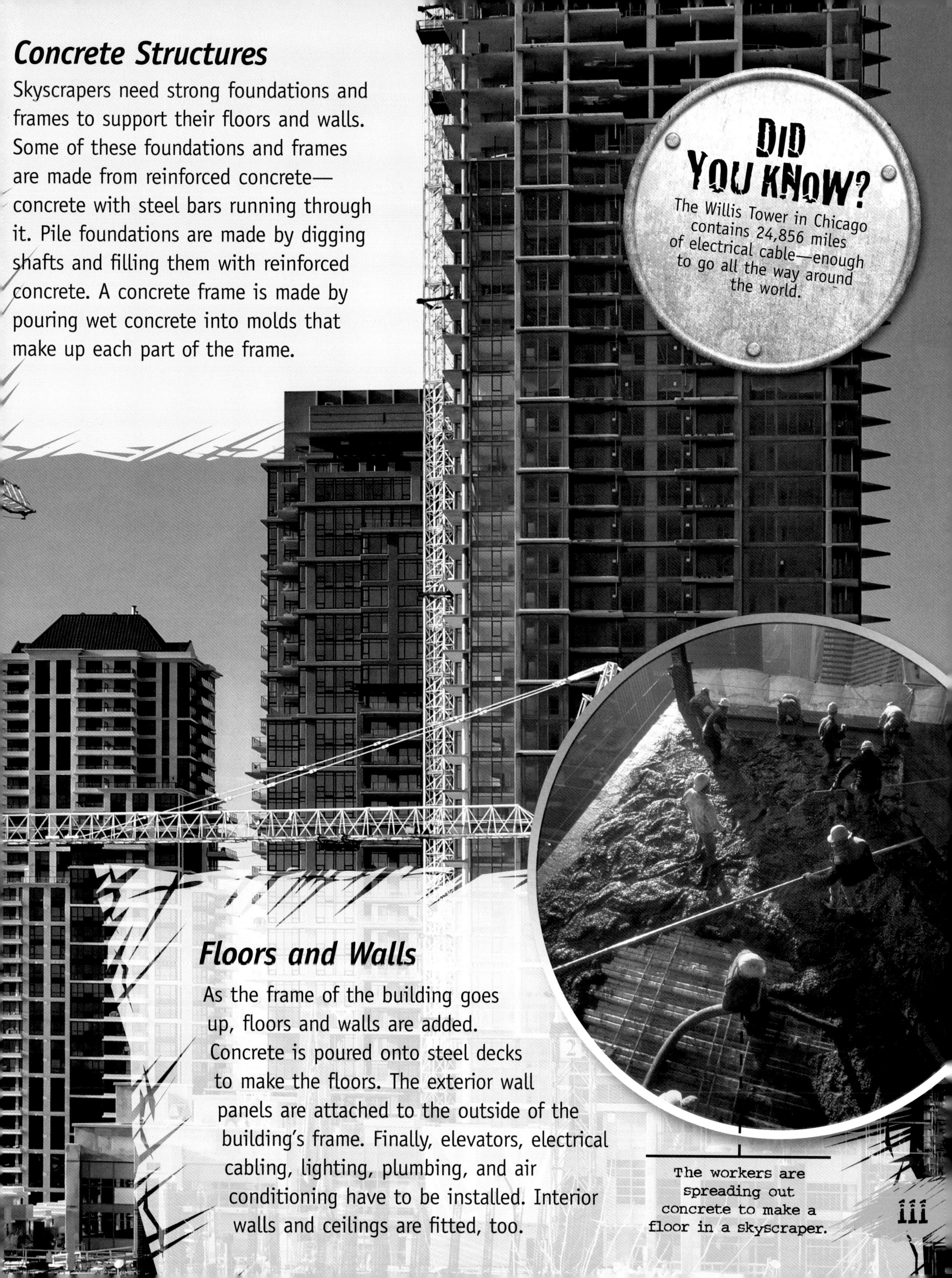

DID YOU KNOW?

The Willis Tower in Chicago contains 24,856 miles of electrical cable—enough to go all the way around the world.

Floors and Walls

As the frame of the building goes up, floors and walls are added. Concrete is poured onto steel decks to make the floors. The exterior wall panels are attached to the outside of the building's frame. Finally, elevators, electrical cabling, lighting, plumbing, and air conditioning have to be installed. Interior walls and ceilings are fitted, too.

The workers are spreading out concrete to make a floor in a skyscraper.

STANDING UP TO NATURE

In some parts of the world, skyscrapers are designed to survive the violent shaking from earthquakes, or the blast of hurricane-strength winds. They are normally given a super strong frame, but some skyscrapers have special features that stop them from swaying.

DID YOU KNOW?
The John Hancock Tower in Chicago swayed up to three feet from side to side in high winds before a tuned mass damper was added.

Earthquake Resistance

Many skyscrapers built in earthquake zones have diagonal braces in their frames to stop the frame from buckling if an earthquake strikes. Other buildings have springs or rubber pads between the frame and the foundations, which reduce the shaking that goes into the building from the ground.

Ground Floor Bracing

The Transamerica Building in San Francisco has diagonal bracing at its base. If an earthquake ever hit San Francisco, this strong bracing would support the building as the ground shook from side to side, and up and down, underneath it.

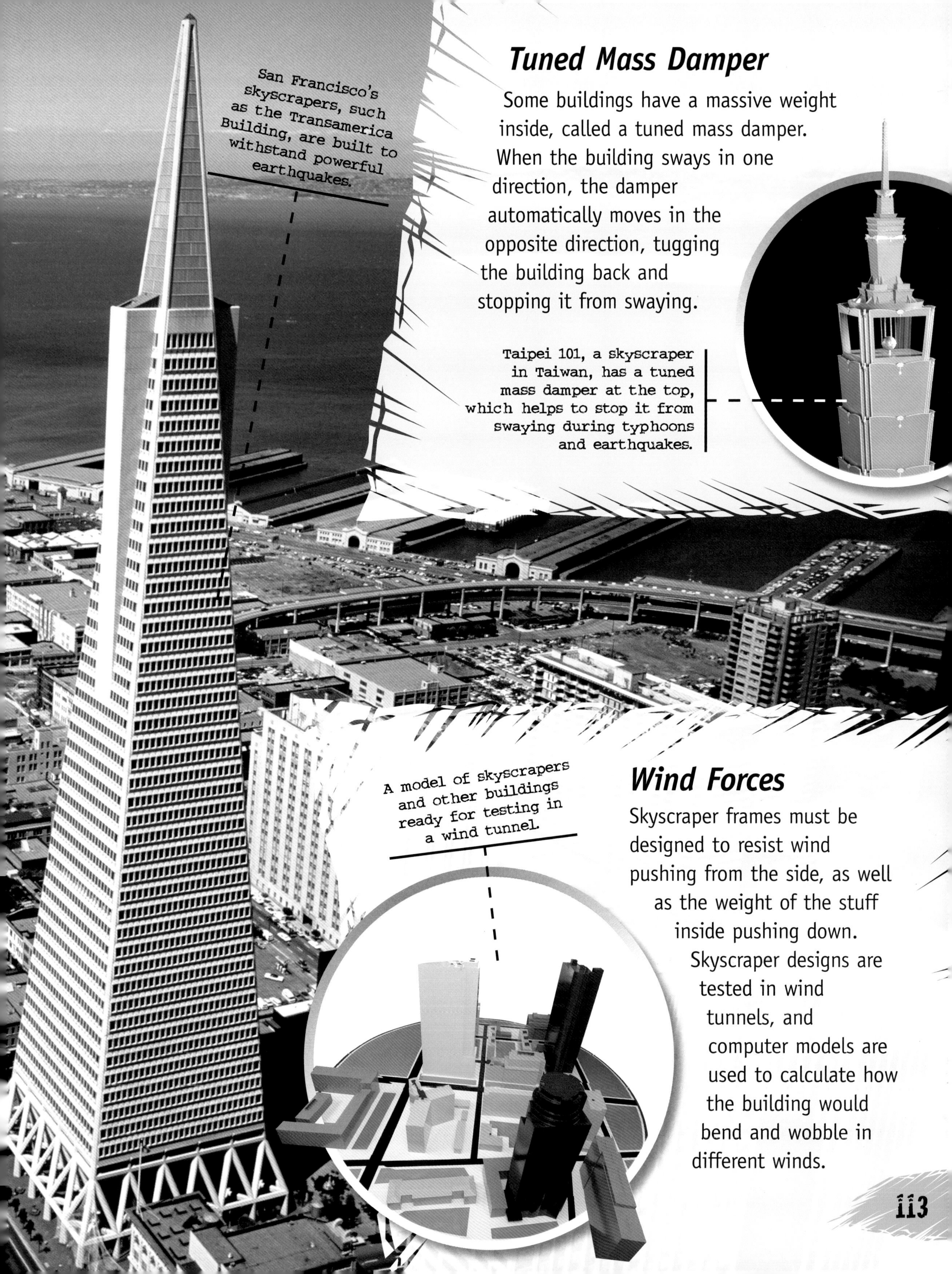

San Francisco's skyscrapers, such as the Transamerica Building, are built to withstand powerful earthquakes.

Tuned Mass Damper

Some buildings have a massive weight inside, called a tuned mass damper. When the building sways in one direction, the damper automatically moves in the opposite direction, tugging the building back and stopping it from swaying.

Taipei 101, a skyscraper in Taiwan, has a tuned mass damper at the top, which helps to stop it from swaying during typhoons and earthquakes.

A model of skyscrapers and other buildings ready for testing in a wind tunnel.

Wind Forces

Skyscraper frames must be designed to resist wind pushing from the side, as well as the weight of the stuff inside pushing down. Skyscraper designs are tested in wind tunnels, and computer models are used to calculate how the building would bend and wobble in different winds.

ELEVATORS

Skyscrapers and other tall buildings would be difficult to use without elevators to carry people to the upper floors and back down again. Elevators are also vital in mines for taking miners deep underground, and in warehouses for moving goods between floors.

Rails and Control

An elevator car is what you stand in. The car moves up and down guided by vertical rails, either fixed in a shaft inside a building, or fixed on the outside of a building. The car normally hangs from a steel cable. An automatic control system moves the car between floors, as people press the elevator-call buttons, keeping track of the car's position and stopping it exactly in the right place for each floor.

The interior of a modern building with an elevator.

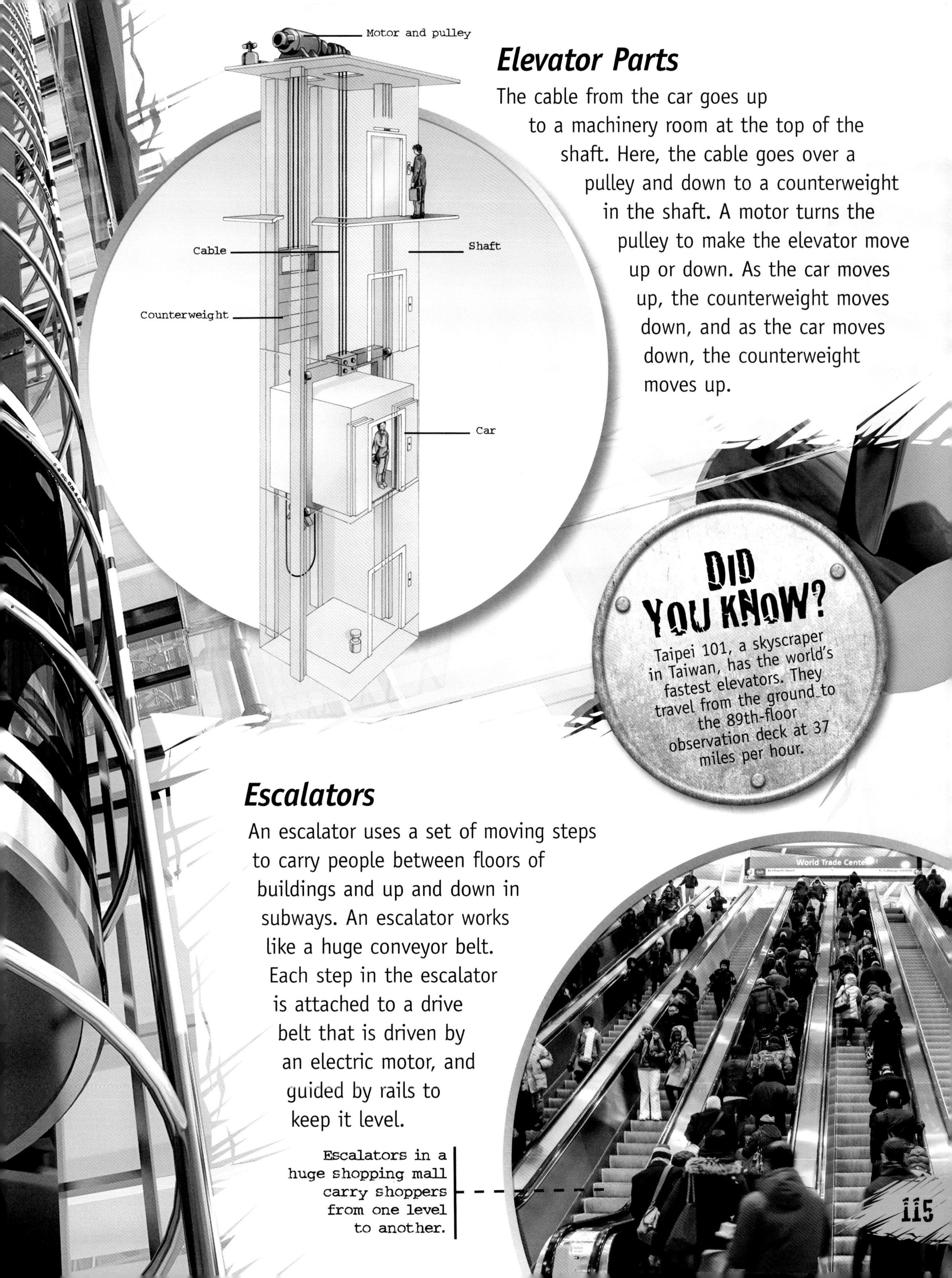

Elevator Parts

The cable from the car goes up to a machinery room at the top of the shaft. Here, the cable goes over a pulley and down to a counterweight in the shaft. A motor turns the pulley to make the elevator move up or down. As the car moves up, the counterweight moves down, and as the car moves down, the counterweight moves up.

DID YOU KNOW?

Taipei 101, a skyscraper in Taiwan, has the world's fastest elevators. They travel from the ground to the 89th-floor observation deck at 37 miles per hour.

Escalators

An escalator uses a set of moving steps to carry people between floors of buildings and up and down in subways. An escalator works like a huge conveyor belt. Each step in the escalator is attached to a drive belt that is driven by an electric motor, and guided by rails to keep it level.

Escalators in a huge shopping mall carry shoppers from one level to another.

CONSTRUCTION MACHINES

Construction machines such as excavators, cranes, earth movers, and dump trucks work on construction sites. They dig holes, lift and carry materials to where they are needed, and move earth and rubble.

This mobile crane has a telescopic boom lifted by a hydraulic ram.

DID YOU KNOW?

The world's tallest mobile crane is the Liebherr LTM 11200, which can lift a load of 1,200 tons 630 feet into the air.

Reaching High

A mobile crane has a boom which is telescopic so that the driver can extend it to reach high into the air. A lifting hook hangs on a cable from the end of the boom. When the boom is extended, legs are also extended from the crane's base to prevent the crane from toppling over.

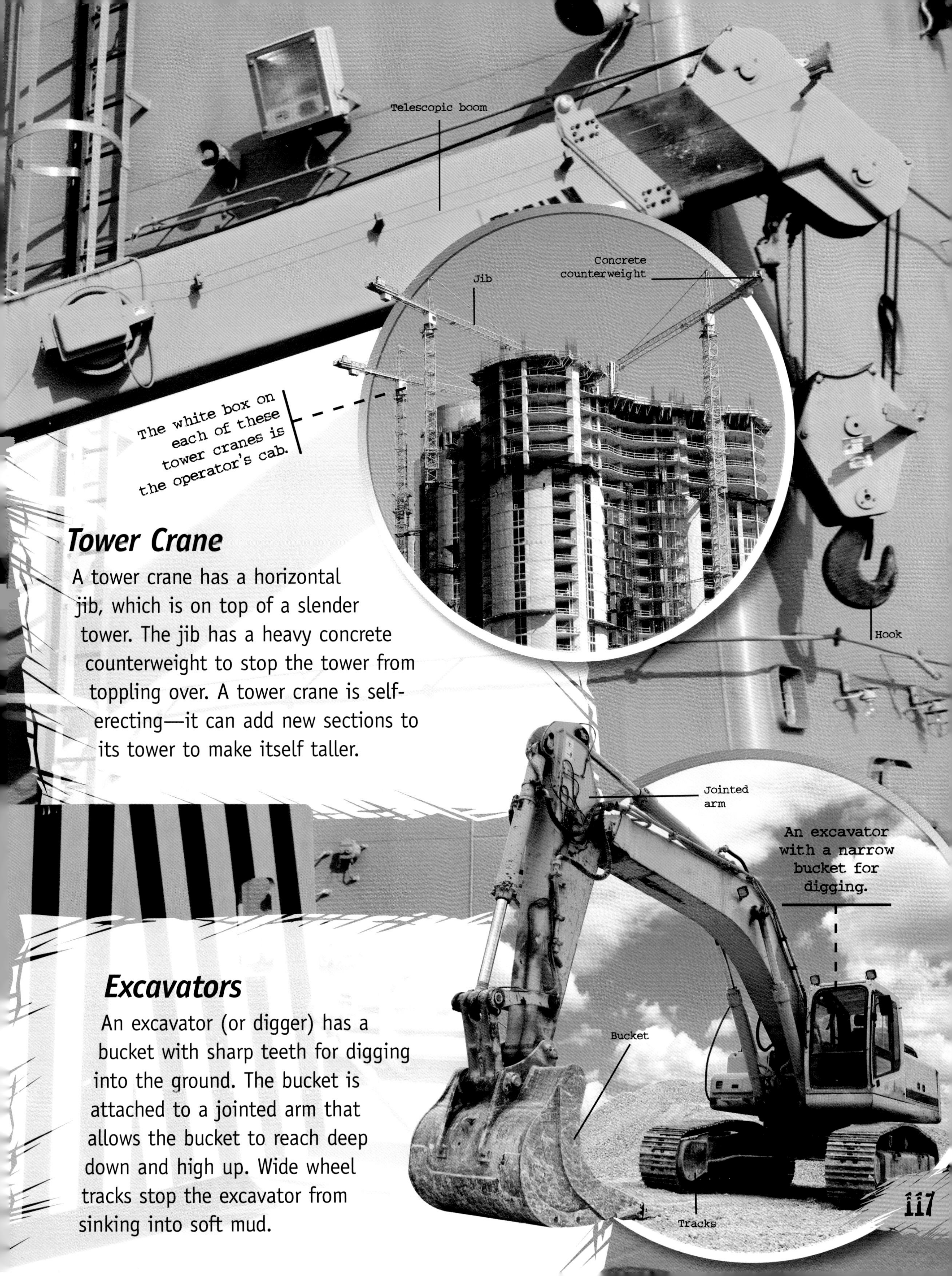
Telescopic boom
Jib
Concrete
counterweight
The white box on
each of these
tower cranes is
the operator's cab.
Hook
Jointed
arm
An excavator
with a narrow
bucket for
digging.
Bucket
Tracks

Tower Crane

A tower crane has a horizontal jib, which is on top of a slender tower. The jib has a heavy concrete counterweight to stop the tower from toppling over. A tower crane is self-erecting—it can add new sections to its tower to make itself taller.

Excavators

An excavator (or digger) has a bucket with sharp teeth for digging into the ground. The bucket is attached to a jointed arm that allows the bucket to reach deep down and high up. Wide wheel tracks stop the excavator from sinking into soft mud.

TUNNELS AND TUNNELING

A tunnel is an underground passage built for roads, railways, water supply, and sewers. Most long tunnels that go deep underground are built by huge tunnel boring machines (TBMs). These machines are as wide as the tunnels they dig.

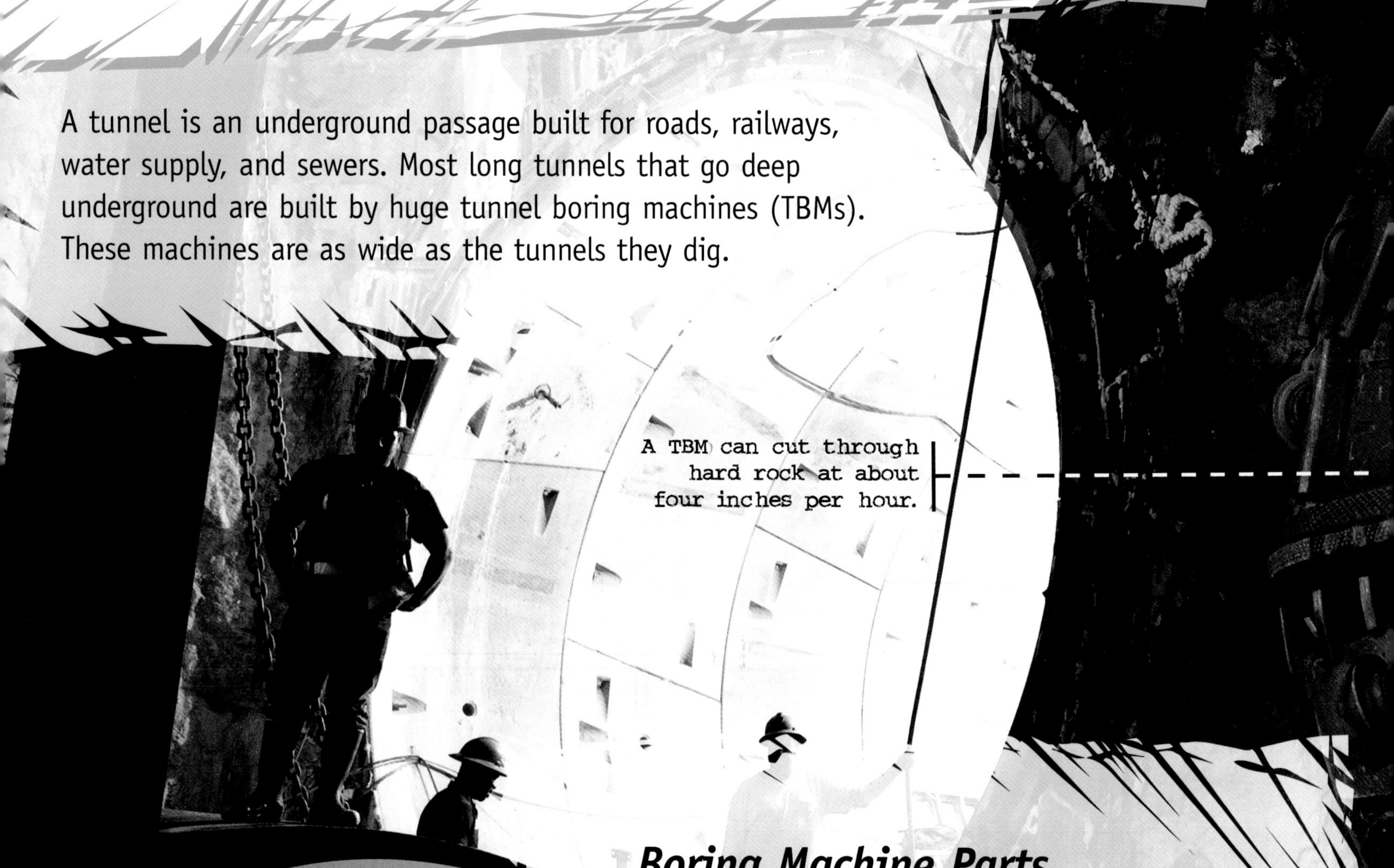

A TBM can cut through hard rock at about four inches per hour.

Boring Machine Parts

A tunnel boring machine is shaped like a cylinder lying on its side. The cylinder, called a shield, supports the rock as the machine moves forward. It has a cutter head covered with rows of tough teeth at the front to cut through the rock. A conveyor carries the broken rock, called spoil, from the cutter to the back of the TBM. Behind the shield is a machine that builds a concrete lining in the new tunnel.

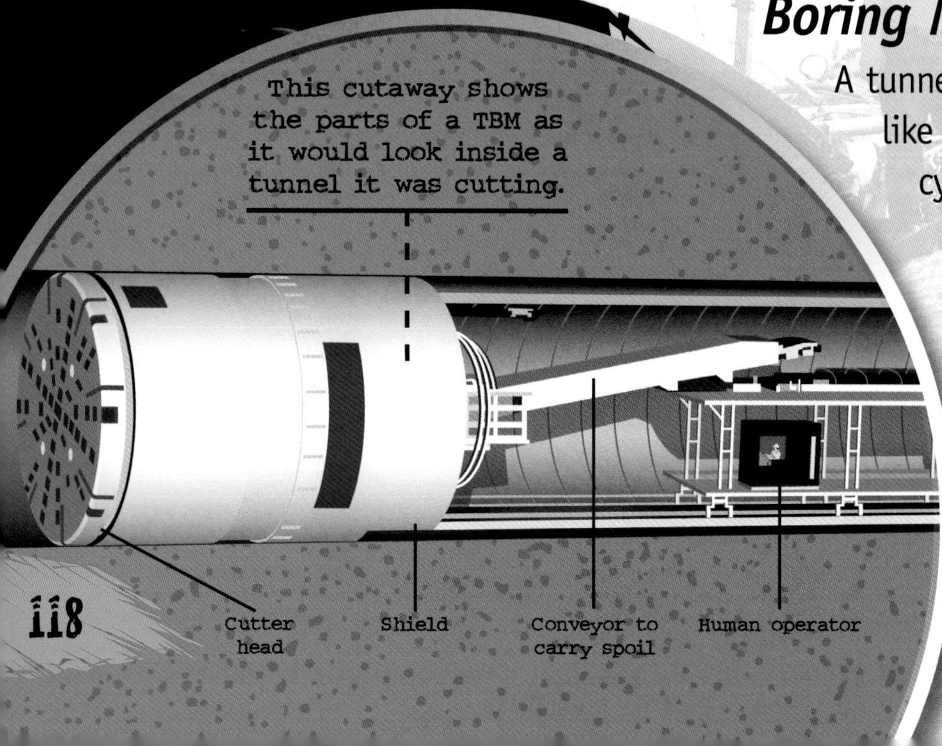

This cutaway shows the parts of a TBM as it would look inside a tunnel it was cutting.

DID YOU KNOW?

The biggest TBMs are almost 52 feet in diameter. With all the machinery that trails behind them, they can be 820 feet long.

Drilling and Blasting

Large tunnels needed in very tough rocks are made using explosives instead of tunnel boring machines. A machine called a drilling jumbo drills holes in the rock. The holes are filled with explosives then the workers move to a safe distance and the rock is blasted out. This process is repeated again and again.

A drilling jumbo makes holes in the rock where the explosives will be placed.

SPORTS STADIUMS

A stadium is a sports ground surrounded by seats for spectators. Like all large structures, a stadium has a strong steel or concrete frame to support the floors and seating areas, and foundations under the ground that support the frame.

DID YOU KNOW?
The roof of the AT&T Stadium in Arlington, Texas, is the world's largest dome. The arches that support the dome are a quarter mile long.

Stadium Parts

The AT&T Stadium in Arlington, Texas, has seats for 80,000 spectators. Tiered seating means that the stadium can hold thousands of spectators, and that everyone can see the action. Wide stairways and corridors behind the seating allow the stadium to fill and empty quickly. The roof is a huge dome supported by steel arches, with a retractable center section.

The AT&T Stadium is the home of the Dallas Cowboys and is the third largest stadium in the National Football League.

Arch Roof Supports

Wembley Stadium in London, United Kingdom, has a capacity of 90,000 spectators. The retractable roof is supported by a steel arch that reaches high above the stadium. Steel cables from the roof are attached to the arch.

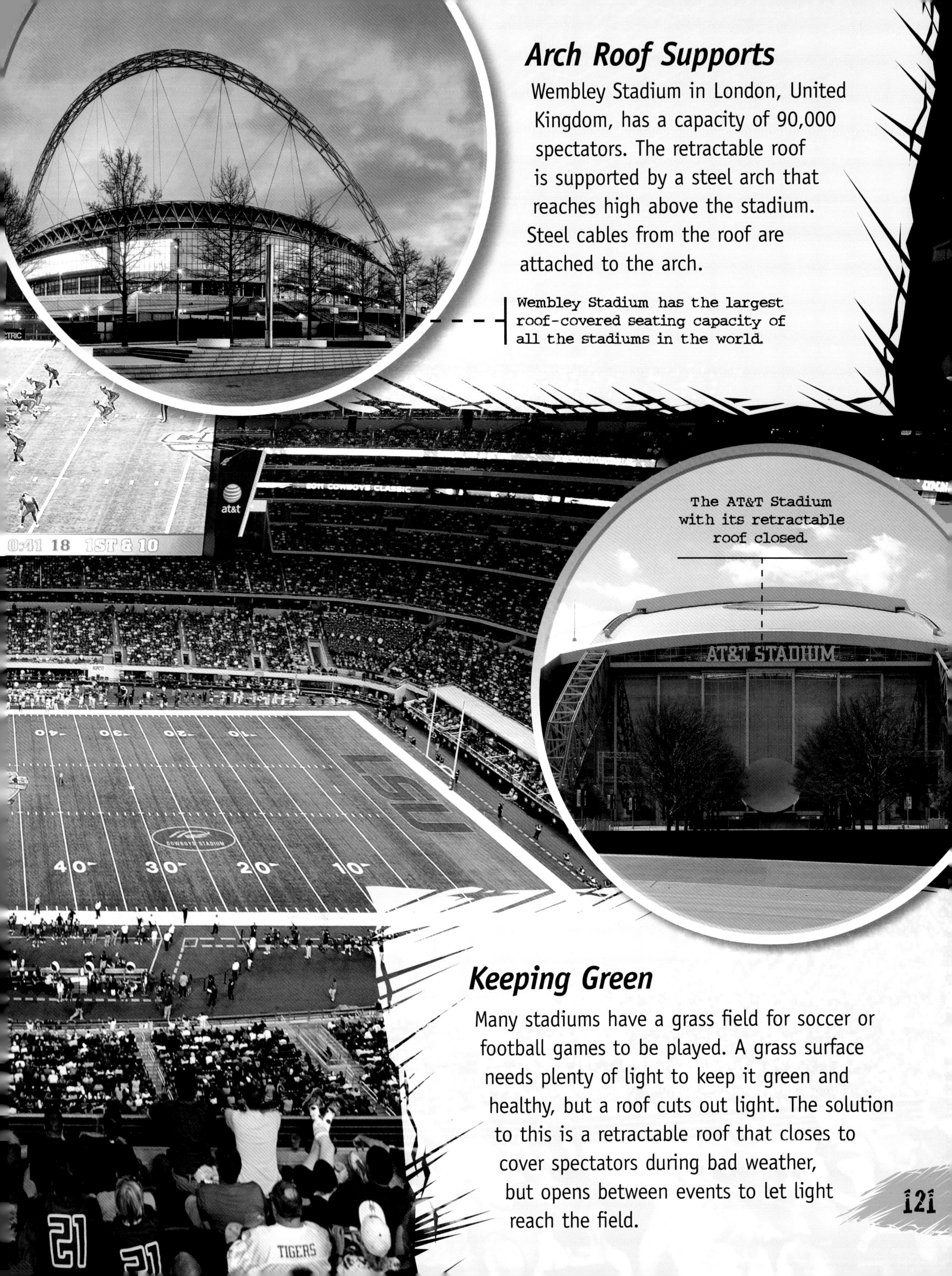

Wembley Stadium has the largest roof-covered seating capacity of all the stadiums in the world.

The AT&T Stadium with its retractable roof closed.

Keeping Green

Many stadiums have a grass field for soccer or football games to be played. A grass surface needs plenty of light to keep it green and healthy, but a roof cuts out light. The solution to this is a retractable roof that closes to cover spectators during bad weather, but opens between events to let light reach the field.

TYPES OF BRIDGES

Bridges are built so that we can cross areas such as valleys, lakes, and rivers. The type of bridge that **engineers** build depends on how much land or water the bridge has to cross, and how long or how high the bridge needs to be. Usually, a beam bridge is used for short bridges and a suspension or cable-stayed bridge is used for long bridges.

The Golden Gate Bridge in San Francisco, California, has about 81 miles of wire inside its suspension cables.

Suspension Bridge

On a suspension bridge, there are suspension cables attached to the tops of the towers. These cables hang down in curves. Vertical cables hang down from the suspension cables. The deck hangs from the vertical cables, called hangers. Most of the world's longest bridges are suspension bridges.

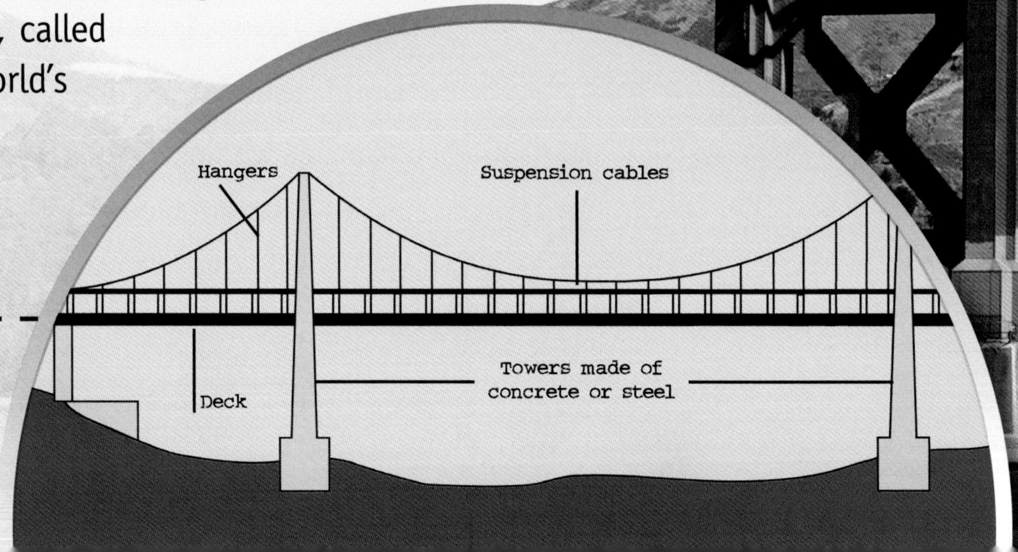

On a suspension bridge, the deck hangs, or suspends, from suspension cables.

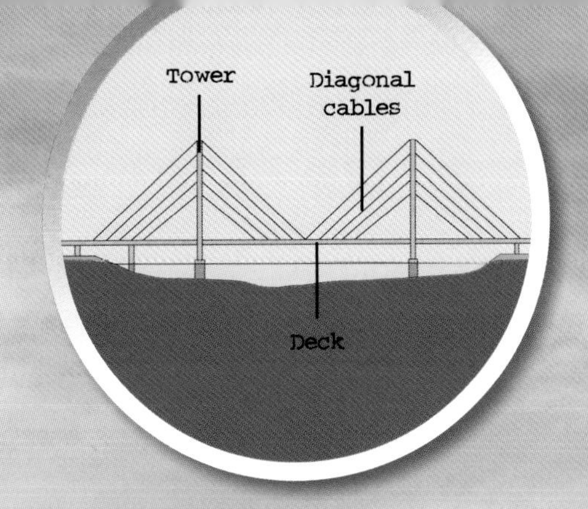

Cable-stayed Bridge

In a cable-stayed bridge, the deck is held up by cables that lead diagonally from the towers to the deck. The thick cables are made of steel and are extremely strong.

Arch Bridge

An arch bridge has a curved shape that rises in the middle. The curved shape of the arch gives the bridge its strength. It carries the weight of the traffic on the bridge to the ground at each end of the bridge.

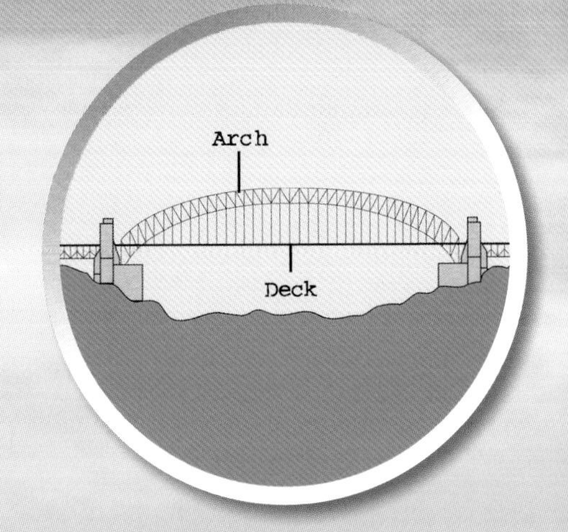

Cantilever beams

Cantilever Bridge

A cantilever is a beam that is held up at one end only. Many cantilever bridges have two cantilevers, one sticking out from each side and meeting in the middle to support the central section.

Beam Bridge

A beam bridge is the simplest type of bridge. It is made up of a strong beam made of concrete or steel, held up at each end.

Beam

DID YOU KNOW?

On very long suspension bridges, the towers at each end of the bridge lean away from each other because of the curve of Earth.

BUILDING BRIDGES

A bridge is a structure that supports traffic, trains, and people as they cross from one side to the other. The long, flat part of the bridge is called a deck. The structure of the bridge holds up the deck. The type of structure used depends on the type of bridge being built (see page 122-123).

Resisting Forces

The weight of the traffic that crosses a bridge pushes down on the bridge's deck. A bridge works by resisting this push, so the deck doesn't fall down. It transfers the weight of the traffic and the deck down to the ground. For example, on an arch bridge, the arch supports the deck from underneath, and the arch presses down on the ground at each end. Long bridges must be designed to stand up to strong winds blowing from the side, too.

DID YOU KNOW?
The Mike O'Callaghan–Pat Tillman Memorial Bridge is the second-highest bridge in the USA and the world's highest concrete arch bridge.
The Mike O'Callaghan–Pat Tillman Memorial Bridge in Colarado was constructed from both sides of the river so the arches, supported by diagonal cables, could meet in the middle.

Opening and Closing

Bridges often have sections that lift up, tilt up, or swing open to let large ships pass through. The 17th Street Causeway Bridge in Fort Lauderdale, Florida, makes room for ships by lifting its deck. This sort of bridge is called a bascule bridge.

Normally a bascule bridge is closed so that traffic can drive across.

When a ship approaches, the beams of the bridge are raised by motors to let the ship pass.

DAMS

A dam is a huge wall built across a river to make a reservoir, which is like a large lake. Water from the reservoir is used in homes, factories, for watering crops, and to help create **hydroelectricity**. Pipes carry the water to a powerhouse, usually behind the dam. The rushing water is used to push turbines which turn a generator to make electricity.

Embankment Dam

An embankment dam is a low dam with gently sloping sides. It is built from earth, rock, and gravel. An embankment dam is built for reservoirs where the water is quite shallow. A waterproof section in the center stops water seeping out.

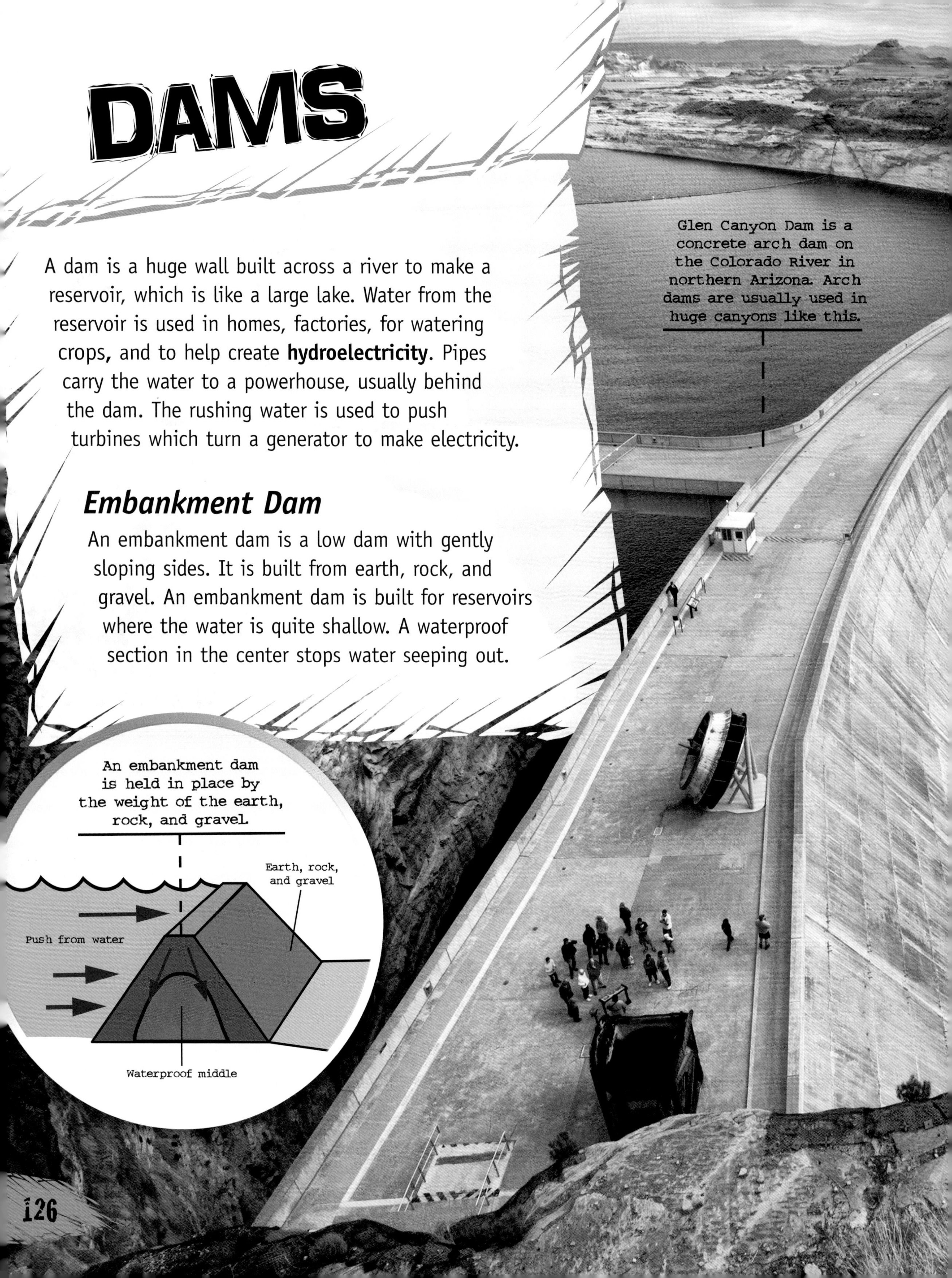

Glen Canyon Dam is a concrete arch dam on the Colorado River in northern Arizona. Arch dams are usually used in huge canyons like this.

An embankment dam is held in place by the weight of the earth, rock, and gravel.

The wall of a gravity dam is thicker at the bottom than at the top.
Concrete
Push from water
DID YOU KNOW?
The Glen Canyon Dam is 709 feet high. It took 400,000 twenty-ton buckets of concrete to build it!
An arch dam transfers the push of the water along the dam into the rock at either end.
Rock
Push from water

Gravity Dam

A gravity dam is a solid concrete wall. Gravity dams are built for reservoirs that are very deep. The huge weight of the concrete wall prevents the powerful push of the water against the dam from knocking it over.

Arch Dam

An arch dam is a curved wall made of reinforced concrete. The wall of an arch dam is thinner than the wall of a gravity dam. Arch dams are used in rocky places such as deep, narrow canyons. The sides and bottom of the dam are fixed to the rock. The arch or curve shape of the wall holds back the water.

POWER STATIONS

At this coal-burning power station, tall chimneys carry steam from the turbines into the air.

A power station produces electricity for use in homes, offices, factories, railways, and many other places. Most power stations produce electricity by burning fossil fuels (coal, oil, and gas). They are called thermal power stations because they use the heat from the burning fuels to produce electricity.

Generating Stages

A power station works by turning the energy from fuel into electricity. The fuel is burned in a furnace to make heat, which boils water in a boiler to make steam. The steam rushes through a **turbine**, which is like a large fan, making the turbine spin. The turbine turns an electricity generator, which produces the electricity.

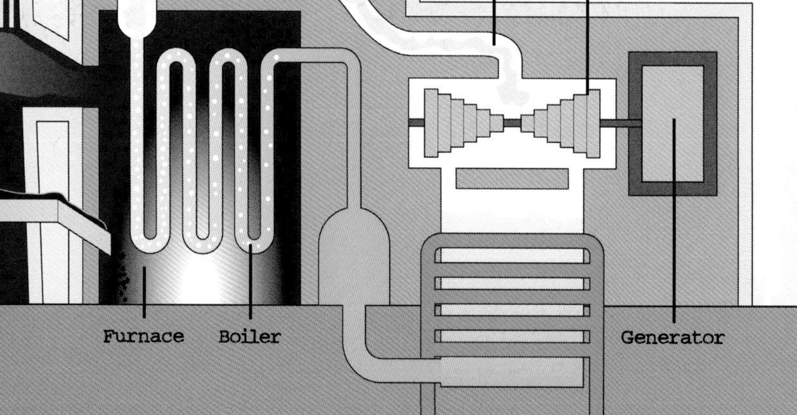

Turbine Hall

In the turbine hall of a power station there are many turbines that are turned by steam. These are linked to the huge generators that produce electricity. Every generator produces at least enough electricity for a small city.

A row of turbines and generators in a power station.

DID YOU KNOW?

The USA's biggest coal-burning power station is in Georgia, and burns 1,288 tons of coal every hour—that's a truckful every minute.

Biomass Power

Biomass is the name given to organic material that is burnt as fuel. Biomass can be logs, wood chips, straw, or any other dried plant material. Biomass is burned in some power stations instead of fossil fuels, or mixed with fossil fuels. Using biomass fuel helps the environment because it produces less carbon dioxide than fossil fuels.

This power plant is using wood chip fuel to create electricity.

SOLAR ENERGY

Solar energy is energy that comes from the sun. Solar energy comes in two forms: light energy, which we can see, and heat energy, which we can't see, but which we can feel on our skin on sunny days. Solar energy is a form of renewable energy.

Photovoltaic cells at the University of New Mexico, Taos campus. Over 2,700 photovoltaic panels will generate enough power for the campus.

Capturing Light Energy

We capture light energy from the sun with photovoltaic cells known as solar cells. These cells turn light from the sun straight into electricity. Toys, watches, calculators, and garden lights often have solar cells to recharge their batteries.

The solar cells on the top of this garden light produce electricity in the day that runs the light at night.

Energy from Light

At a solar photovoltaic power station, thousands of photovoltaic cells work together to produce large amounts of electricity. The banks of cells rotate as the day passes, so that they always face the sun. Solar power stations are built in parts of the world that have a lot of sunshine.

Heat Energy

Solar thermal energy is heat energy we catch from the sun. At a solar thermal power station, the heat is focused using mirrors. Then, the heat is used to boil water to make steam, which drives turbines, which then spin generators to produce electricity. Some homes have solar water heaters that catch solar thermal energy used to heat water for washing and showering.

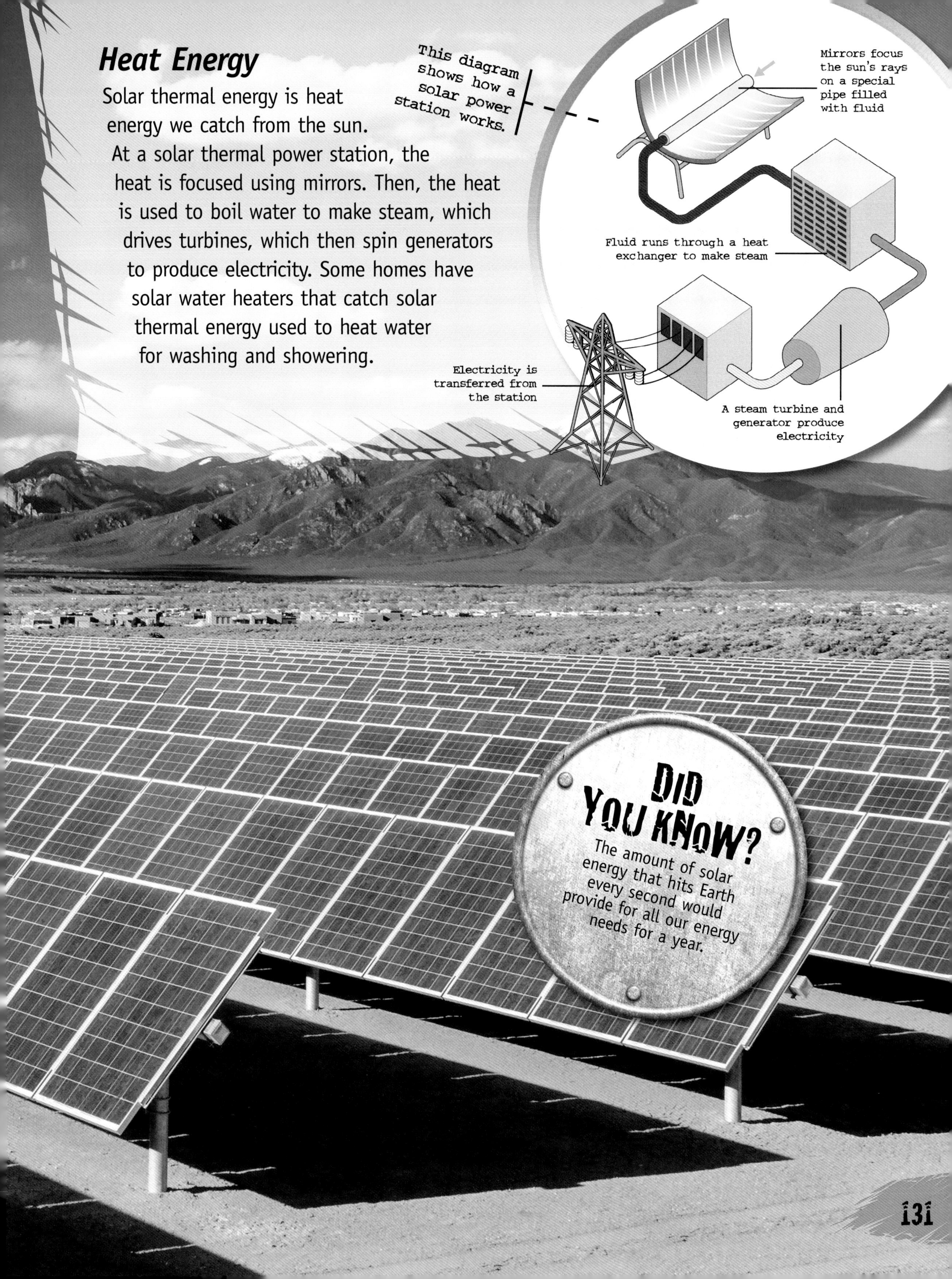

This diagram shows how a solar power station works.
Mirrors focus the sun's rays on a special pipe filled with fluid
Fluid runs through a heat exchanger to make steam
Electricity is transferred from the station
A steam turbine and generator produce electricity
DID YOU KNOW?
The amount of solar energy that hits Earth every second would provide for all our energy needs for a year.

WIND AND WATER ENERGY

Some forms of renewable energy come from the natural movements of air and water. We capture the energy in wind, the energy in water flowing down rivers, and the energy in tides and waves.

DID YOU KNOW?

The world's largest wind turbine is the Vestas V164. Each of its three blades is 262 feet long.

Wind Farms

We capture wind energy with wind turbines. A wind farm is made up of tens or hundreds of wind turbines. Wind farms are built in places where it's windy a lot of the time. They are often built out at sea, where the winds are normally stronger than on land.

Wind Energy

A wind turbine is made up of a rotor, which is held by a control system called a nacelle on top of a tower. The wind makes the rotor spin, and the rotor turns an electricity generator inside the nacelle. The nacelle rotates so that the rotor faces into the wind all the time.

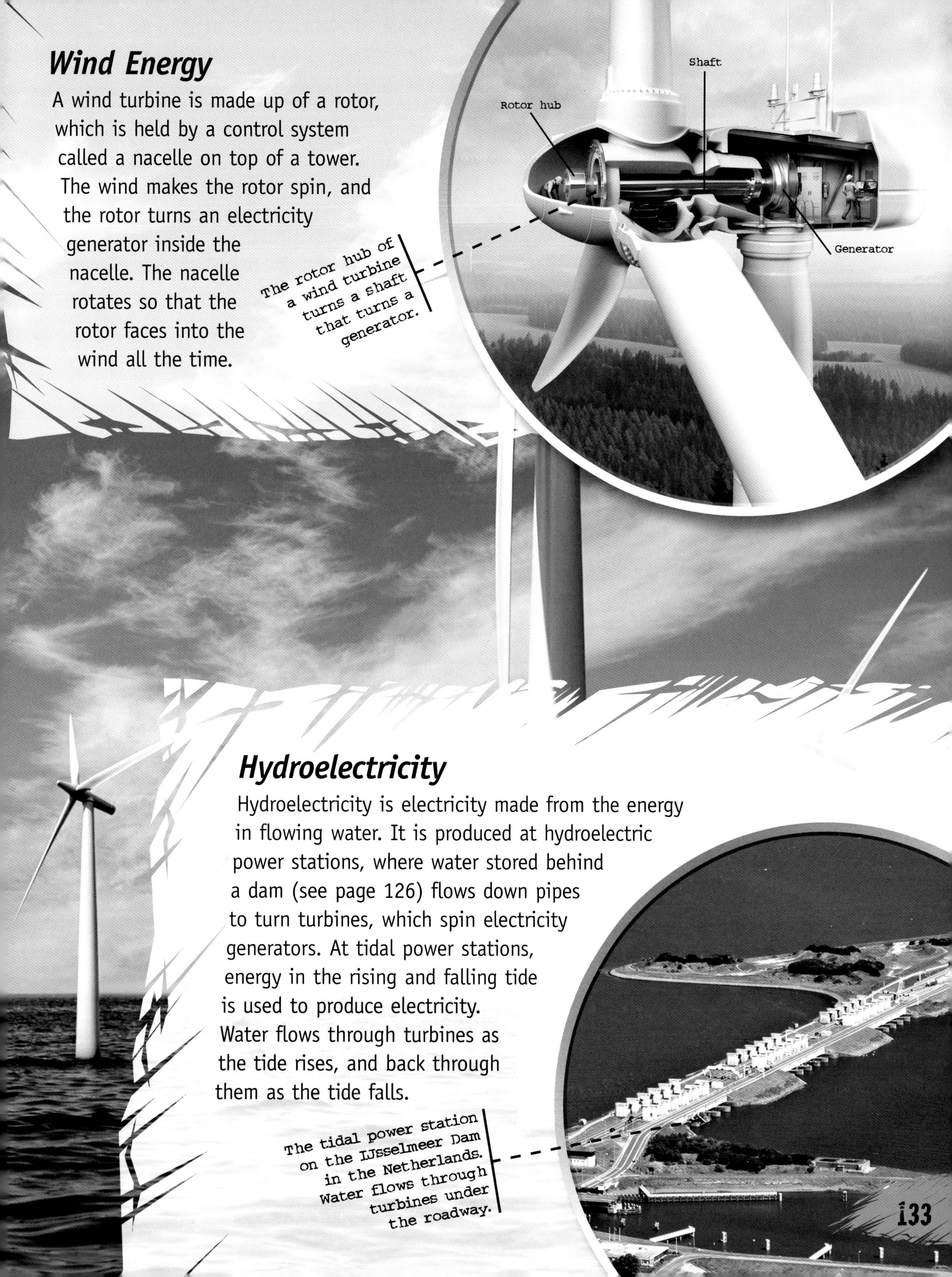

The rotor hub of a wind turbine turns a shaft that turns a generator.

Hydroelectricity

Hydroelectricity is electricity made from the energy in flowing water. It is produced at hydroelectric power stations, where water stored behind a dam (see page 126) flows down pipes to turn turbines, which spin electricity generators. At tidal power stations, energy in the rising and falling tide is used to produce electricity. Water flows through turbines as the tide rises, and back through them as the tide falls.

The tidal power station on the IJsselmeer Dam in the Netherlands. Water flows through turbines under the roadway.

NUCLEAR ENERGY

The giant cooling towers at a nuclear power plant cool the water from the condenser before it is returned to a lake.

Nuclear energy is contained in atoms, which are the tiny particles all substances are made from. When the central part of an atom, called the nucleus, splits up into two pieces, some of this energy is released. This is called nuclear fission. In nuclear power stations, this energy is used to produce electricity.

Parts of a Nuclear Power Station

In a nuclear power station, heat is produced through nuclear fission in a reactor. The heat boils water to make steam, and the steam turns turbines, which spin generators to produce electricity. In the condenser, water from a lake, river, or the sea, is used to cool the steam to turn it back to water. Nuclear reactors release dangerous rays, called radiation. A reactor is enclosed in a thick concrete chamber to prevent this radiation from leaking out.

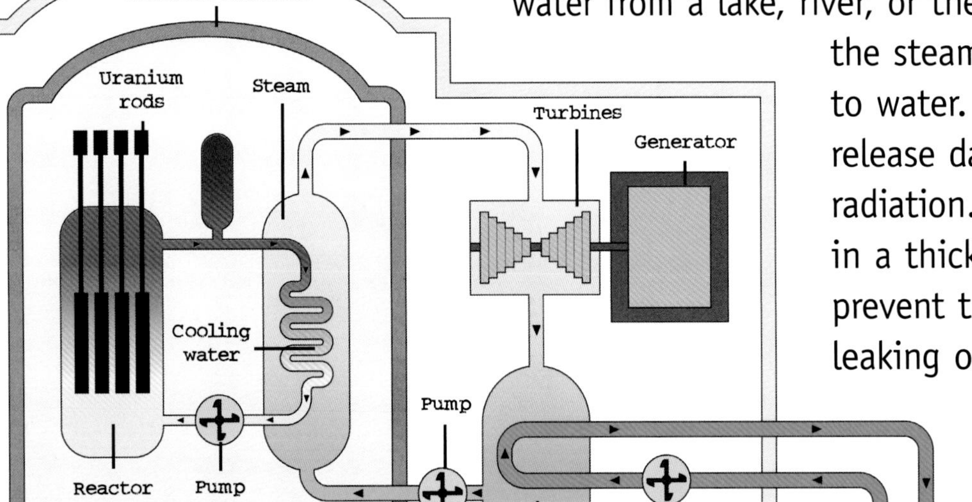

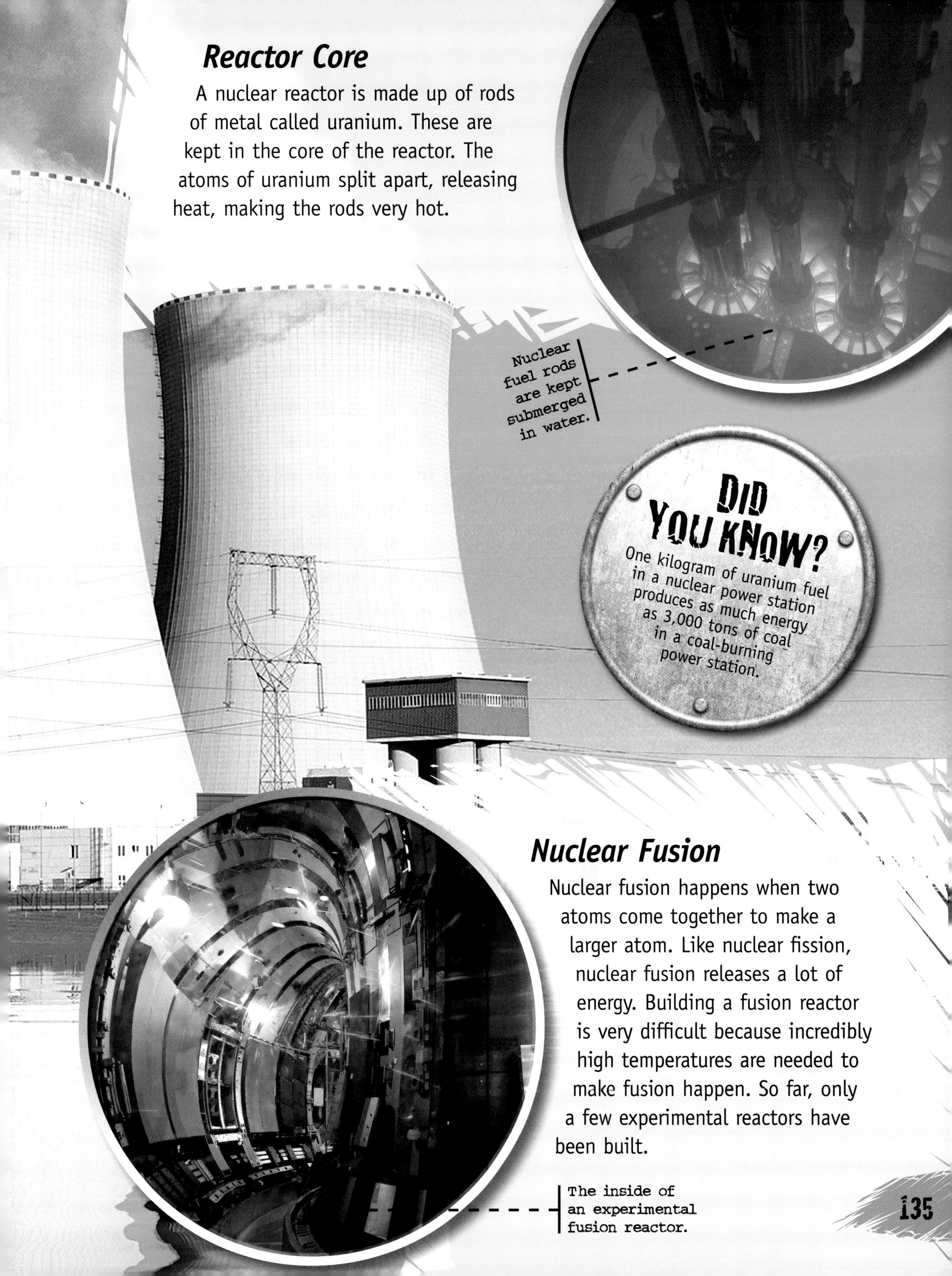

Reactor Core

A nuclear reactor is made up of rods of metal called uranium. These are kept in the core of the reactor. The atoms of uranium split apart, releasing heat, making the rods very hot.

Nuclear fuel rods are kept submerged in water.

DID YOU KNOW?
One kilogram of uranium fuel in a nuclear power station produces as much energy as 3,000 tons of coal in a coal-burning power station.

Nuclear Fusion

Nuclear fusion happens when two atoms come together to make a larger atom. Like nuclear fission, nuclear fusion releases a lot of energy. Building a fusion reactor is very difficult because incredibly high temperatures are needed to make fusion happen. So far, only a few experimental reactors have been built.

The inside of an experimental fusion reactor.

DRILLING FOR OIL AND GAS

Oil and gas are trapped in layers of rock under the ground. They are brought up to the surface by drilling deep holes called wells and allowing the oil and gas to flow up through pipes.

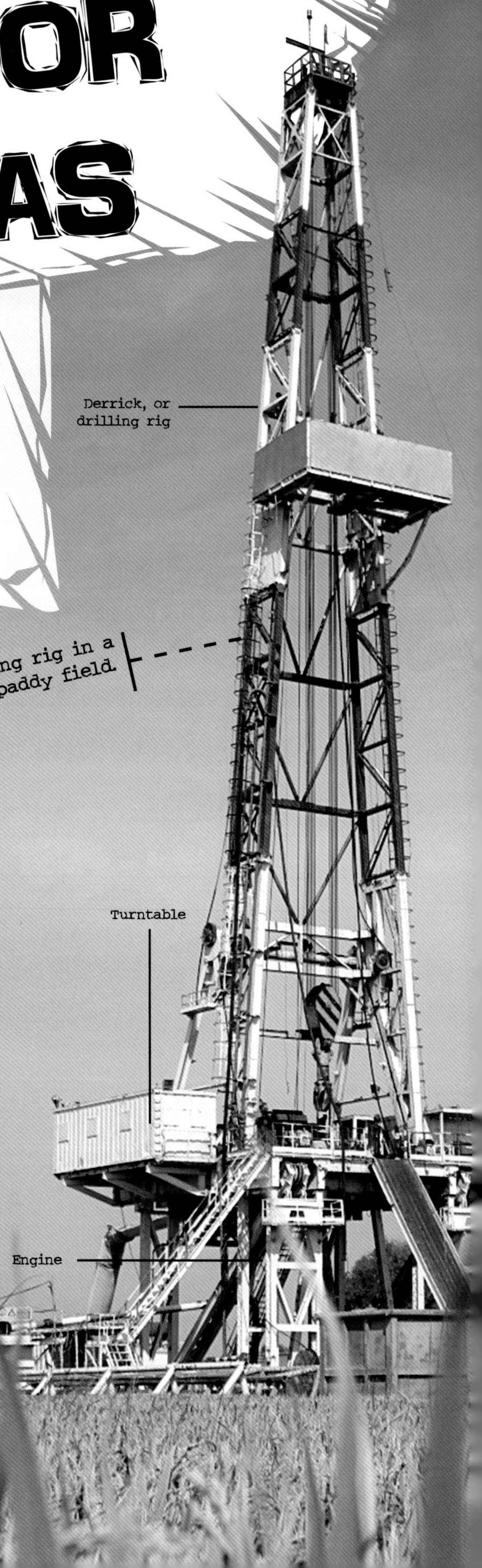

Derrick, or drilling rig

A land drilling rig in a Chinese paddy field

Drilling Rig

The drill is a metal pipe with a drill bit at the end. The drill is held by a structure called a derrick, or drilling rig. An engine turns the drill, and the drill pushes down into the rock to make a hole or well. Pipes are put into the well, and as the well gets deeper, more lengths of pipe are added. The oil or gas flows through the pipes from under the ground to the surface. Once the drill strikes oil or gas, the top of the well is fitted with valves to control the flow.

Turntable

Engine

Steering a Well

Drills do not have to go straight down. The end of a drill can be aimed in a different direction. This is called directional drilling. It is used where engineers have to drill sideways from the top of the well to get to the oil or gas.

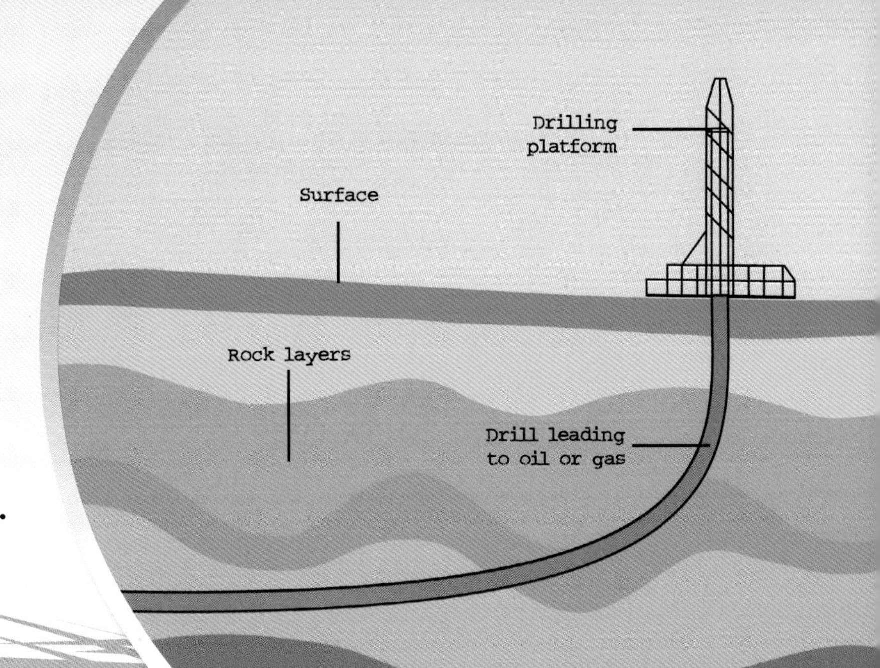

DID YOU KNOW?

Oil was first successfully drilled in Titusville, Pennsylvania, in 1859. In 2015, U.S. oil production is about 9.6 million barrels a day.

A pumpjack pumping oil up out of an oil well in Colorado.

Pumping Oil

Once oil has been found by drilling a well, the oil is pumped to the surface through pipes. Sometimes the pump is lowered into the well, and sometimes it sits at the top of the well. A pumpjack, also a called a "nodding donkey," has a large lever arm that rocks back and forth, operating a pump similar to a bicycle pump inside the well.

OFFSHORE DRILLING

Offshore platforms are drilling rigs that work at sea to search for, and collect, oil and gas from rocks deep under the seabed. The platforms are giant structures that support a drilling rig, drilling equipment, and places for the workers to eat, relax, and sleep above the waves. Some platforms are as tall as skyscrapers.

The Eirik Raude platform moves from place to place and can drill in water up to 9,842 feet deep.

Semi-submersible Platforms

Some platforms float on the surface of the sea. They are semi-submersible, which means there are enormous ballast tanks full of water under the surface to make them stable in waves. They are anchored to the seabed by cables to stop them drifting out of position. Some semi-submerisble rigs have their own engines for moving between drilling sites.

Towing Rigs to Sea

Semi-submersible platforms and fixed platforms are built in the water, close to land. When a platform is complete, it is towed out to sea to where it is going to drill. The legs of fixed platforms are then lowered until they touch the seabed. The ballast tanks of semi-submersible platforms are then flooded to lower the platform into the water.

Tugboats tow the platform to the drilling site.

DID YOU KNOW?

A concert was held at the base of the Troll A platform to celebrate its 10th anniversary. It set a new world record for the deepest underwater concert at 994 feet below sea level.

The Troll A gas platform stands about 50 miles northwest of Bergen, Norway.

Fixed Platforms

Fixed platforms stand on the seabed on tall legs made of steel or concrete. On some platforms, the legs are hollow tubes, which are full of seawater. The Troll A gas platform's enormous concrete legs are more than 1,200 feet high and weigh an astonishing 656,000 tons.

GLOSSARY

Aerial
A device that transmits or receives radio signals.

Altitude
A measure of height above the ground.

Carbon fiber
A material made of thin threads of carbon mixed with plastic.

Combustion
A chemical reaction that causes fire.

Composite
A material made of two or more other materials, such as carbon fiber which is made from carbon and plastic.

Data
Information that is normally stored on a computer, tablet, or mobile phone.

Diesel engine
An internal combustion engine that uses diesel oil as a fuel.

Engineers
People who design, build and look after machines or buildings.

Exhaust
The part of an engine where gases made from burning fuel escape into the air.

Foundation
The part of a building that is underground and that holds up the rest of the building.

Fuel
A solid, liquid, or gas that is burned in engines or in power stations to make something move or to make heat.

Gears
A set of toothed wheels that make parts of machines turn at different speeds; used in vehicles to move at different speeds.

Generator
A machine that turns the movement of a shaft into electricity.

Horsepower
A unit of energy.

Hybrid
Made up of two different things.

Hydraulic ram
A machine made up of a cylinder and a piston, in which the piston is moved by pumping liquid into or out of the cylinder.

Hydroelectricity
Electricity produced by capturing the energy in flowing water.

Internal combustion engine
Fuel, such as gasoline or diesel, burns in this type of engine.

Kevlar
An artificial material that is extremely strong.

Laser beam
A very narrow beam of intense light.

Latitude
A measurement of how far a place is from Earth's equator.

LED
Short for light-emitting diode, an electronic device that turns electricity into light.

Locomotive
A railway vehicle that provides the power to pull trains.

Longitude
A measurement of how far around Earth a place is, measured from Greenwich in the United Kingdom.

Magnetic field
The area around a magnet where the magnet's force can be felt.

Orbit
The route that a spacecraft takes around Earth.

Processor
A complex electronic circuit that is the brain of a computer, tablet, game machine, or smartphone.

Radar
A system that detects objects by firing radio waves into the air and determining which waves bounce back.

Radiation
Invisible waves or rays given out by radioactive material or by stars in space.

Radio signal
A radio wave that has been shaped to carry information from place to place.

Radio waves
Invisible waves related to light and microwaves.

Range
The distance from a gun to the target it is aiming at.

Receiver
An aerial that detects radio waves.

Reinforced concrete
Concrete with steel bars embedded in it for extra strength.

Solar cell
An electronic device that turns light into electricity.

Streamlined
The smooth shape of a vehicle's body, which allows it to move easily through the air.

Supersonic
Faster than the speed of sound.

Suspension
Springs and other parts that allow the wheels of a vehicle to move up and down as the vehicles goes over bumps.

Tether
A special rope that connects an astronaut to a spacecraft or space station.

Thrusters
Small engines that send out jets of gas in spacecraft or water in ships and other watercrafts.

Transmitter
An aerial that sends out radio waves.

Turbine
A machine like a fan that spins when air, gas, or steam flows through it.

Ventilation
A system that lets in fresh air.

Weld
To join two pieces of metal by melting the joint between them, or to melt two pieces of metal together.

INDEX

Picture Credits

Key: t = top, b = bottom, c = center, l = left, r = right

Alamy 4–5c Design Pics Inc; cb epa european pressphoto agency b.v, 24–25c Kevin Foy, 26–27 qaphotos.com, 38–39b digitalunderwater.com, 48–49b ZUMA Press Inc; tr Malcolm Park editorial, 58–59l Everett Collection Historical; br US Navy Photo, 64–65br bavariaimages, 68–69c NG Images, 78–79tr sciencephotos; bl epa european pressphoto agency b.v, 88–89c Gregg Vignal; l M40S Photos, 102–103c dpa picture alliance archive; tr Universal Images Group Limited, 112–113c Dennis Hallinan, 130–131c Design Pics Inc, 132–133c Simon Belcher, 132–133br frans lemmens.

Corbis 4–5tr PASIEKA/Science Photo Library, 12–13c Vw/dpa/Corbis, 18–19c U.S. Navy—digital version copy/Science Faction, 20–21tr Jon Bower/LOOP IMAGES; bc Noboru Hashimoto/Sygma, 26–27tcr Jacques Langevin/Sygma, 48–49c YONHAP/epa, 58–59c Michael Lewis, 60–61c NASA/Handout/CNP, 68–69tr Elena Duvernay/Stocktrek Images, 76–77b PASIEKA/Science Photo Library, 100–101c Roger Ressmeyer, 110–111br Joel W. Rogers, 118–119c Timothy Fadek, 138–139c Courtesy Ocean Rig/Reuters, 138–139br Statoil.

Dreamstime 8–9tr Luminis, 22–23tr Lawcain, 56–57c Amskad, 80–81bl Tele52, 82–83tr Stihl024, 96–97r Kasinv, 104–105b Artofchris.

Getty 1 Stone, 14–15 tr STAN HONDA/AFP, 18–19 br Scott Nelson/Getty Images, 22–23c Moment Mobile; bc Tom Williams/Roll Call, 28–29b Thomas Koehler/Photothek, 30–31bl Stone, 42–43bl YURI KADOBNOV/AFP; b Christian Bach/ullstein bild, 50–51t Photographer's Choice, br Moment, 56–57r Lonely Planet Images, 60–61br NASA, 62–63tr E+; bl Ricky Carioti/The Washington Post; c World Perspectives, 76–77c Kevork Djansezian, 102–103l Encyclopaedia Britannica/UIG, 106–107b Pier Marco Tacca; t Richard Juilliart/AFP, 114–115t Dorling Kindersley, 120–121c Rob Tringali/SportsChrome, 132–133tr Stone, 134–135b Monty Rakusen.

iStock 6–7br Jonathan Woodcock, 8–9br Jason Doiy, 10–11tc magnetcreative, 28–29tr Hhakim, 36–37c onlymehdi, 116–117r Maxian, br jordan_rusev, 118–119br Eduard Andras, 128–129tr BanksPhotos.

Shutterstock 2–3 Anton Foltin, 4–5lc Yeamake, 6–7bl Paolo Bona; c Kurkul, 8–9c Darren Brode, 10–11c Dan Schreiber; br Asturcon, 12–13tc David Acosta; br Angyalosi Beata, 14–15bl Rob Wilson; c Pierre Desrosiers; br Nadezhda Bolotina, 16–17c Ramon Espelt Photography; br Tyler Olson; tr Herbert Kratky, 18–19tr Sergei Butorin, 20–21c Oleksiy Mark, 22–23bl Paolo Bona, 24–25tr Dmitry Kalinovsky; bc Jaromir Chalabala, 28–29c Rigucci, 30–31tr AnneMS; c nattanan726, 32–33c sainthorant daniel; br PomInOz, 34–35tr nevenm; c Italianvideophotoagency; bc Gustavo Miguel Fernandes; br Yuriy_fx, 38–39c iurii, 40–41tr Tyler Olson; c Margo Harrison, 42–43c topseller, 44–45br Bocman1973; bc Sushkin; c Cherkas, 46–47c travellight; br Mario Hagen; tr Frank Wasserfuehrer, 50–51c Slavoljub Pantelic, 52–53br James A. Harris; t Richard Thornton; c Patrick Wang, 54–55c Burben; tr hxdyl; b Federico Rostagno, 64–65c Andrey Armyagov, 66–67br Rafal Olechowski; c Boris Rabtsevich, 70–71c Mpanchenko; b Szasz-Fabian Jozsef, 72–73t isak55; bl Aleksei Lazukov; br Mrklong; c Alexandr Svistakov, 74–75b Zern Liew; c mama_mia, 78–79c Jordan Tan, 80–81t mama_mia; c Rido, 82–83c domjaves; b Yen Hung, 84–85b Vladimir Arndt; c hin255, 86–87tr Ohmega1982; c PathDoc; cr Yeamake, 88–89b Dusan Jankovic, 90–91c Bacho; r fantom_rd, r sahua d, 92–93c wavebreakmedia; r prochasson frederic, 94–95c Luciano Mortula; cr Anthony Berenyi; t Menna 96–97c supergenijalac; t DenisKlimov, 98–99tr OliverSved; b Monkey Business Images; c Tyler Olson; l eAlisa, 100–101b fstockfoto, 104–105t Designua; c Jorg Hackemann, 106–107 c SSSCCC, 108–109c Migel; l Matej Kastelic, 110–111c Alex533; bl LauraKick, 114–115br iurii; c f11photo, 116–117c Thisislove, 120–121r Katherine Welles; t photo.ua, 122–123c dibrova, 124–125br R. Gino Santa Maria; c Anton Foltin; b Christian Colista, 126–127c Alexey Kamenskiy; 128–129c VanderWolf Images; b MarekPiotrowski, 130–131b PhIIIStudio, 134–135tr Everett Historical; c Kletr, 136–137c zhuda; br rCarner, 138–139t, 140–141 esbobeldijk, 142–143 Francois Roux, 144–144 4Max.

Other (Non agency)

56–57bl ©NASA http://www.nasa.gov,
58-59tr ©NASA http://www.nasa.gov.

Kuo Kang Chen

Illustrations on 39, 41, 64, 67, 71, 81, 82, 92, 100–101, 128, 131 and 134.